THE ENERGY OF
HEBREW LETTERS

*The Energy of the Hebrew Letters is a revised edition of The Power of Aleph Beth Vol 1 and Vol 2.

Rav Berg wishes to acknowledge the contributions of Rabbi Levi Krakovsky to certain of the concepts contained in this book and to scholarship in the field of Kabbalah.

Kabbalah Publishing is a registered DBA of Kabbalah Centre International, Inc.

For further information:

The Kabbalah Centre
155 E. 48th St., New York, NY 10017
1062 S. Robertson Blvd., Los Angeles, CA 90035

1.800.Kabbalah www.kabbalah.com

First Edition
February 2010
Printed in Canada
ISBN13: 978-1-57189-640-7

Design: HL Design (Hyun Min Lee) www.hldesignco.com

Mixed Sources
Product group from well-managed forests, controlled sources and recycled wood or fiber
www.fsc.org Cert no. SW-COC-000952
© 1996 Forest Stewardship Council
FSC

THE ENERGY OF
HEBREW LETTERS

RAV BERG

DEDICATION

To Karen, my dear wife,
who introduced me to
the joys of balancing the world,
with gratitude, admiration and love.

TABLE OF CONTENTS

PREFACE

Can we really learn to control our internal processes with our minds instead of with pills or a soothing glass of something on the rocks? I believe that until we come to grips with the realization that there is more to life than the needs of the physical body, we shall be forced to endure the vicissitudes of a life of uncertainty.

We live in an extraordinary age. All of us must wonder from time to time if there is any real purpose to life. We all have immediate goals; we all want to achieve some gratification in our work and to bring up our offspring. But what of a long-range purpose?

The *Zohar* (the sacred text of Kabbalah) is an acknowledged source of great spiritual wisdom that is as ancient as the Bible itself. The *Zohar* is considered the "decoder" of the Bible. It was passed on as oral tradition until it was recorded as text that remained hidden for thousands of years. The *Zohar* is intended for people of all ages whose imaginations have not been stifled by the standard Cartesian educational process. It is written for people who are not afraid to embrace "new" ideas, even though these ideas may be a thousand years old—or even older.

It is the intention of this book to travel back to the beginnings of the universe and the Big Bang. Understanding the point from which it all began is in the *Zohar*. And because the Bible, according to *Zoharic* interpretation, is a cosmic code awaiting deciphering, the Hebrew letters provide the perfect cosmic bond. It is to this task of decoding that the Kabbalah addresses itself. There are a number of these themes woven through the discussion of the structure and energy of the letters *Tav* to *Lamed*.

In the revelation of the energy of the letters *Kaf* to *Alef*, we seek to answer many vexing questions concerning destiny, fate, and free will. I thought it might be interesting to examine the nature of the universe and our role in it and to explore more closely how the doctrines of the Bible agree with—or contradict—the discoveries of science. The range of topics may seem eclectic—from the extinction of the dinosaurs to the structure of the cosmos—but these topics reflect the essence of the Bible and the cosmic code that lies at the heart of reality, and they provide an overview of the cosmic origins of nature and man.

Within the forthcoming chapters, the reader will be given the opportunity to enter a dimension of reality where he or she will no longer be an unfortunate victim of circumstances, but rather, the creator of his or her own destiny. If humanity is to exist as more than a mere link in the cosmic chain of uncertainty, we shall need to dispense entirely with the illusion that poses as reality in the physical world.

Kabbalah is meant to make us free. It can give us wings with which to explore the world. It can also free us in a more mundane way by alerting us to our unique personal challenges and giving us the conceptual tools to overcome them. The astronaut suspended in space sees the Earth as it truly is: small and blue and beautiful in the eternal silence where it floats. As the astronaut passes from sunlight into darkness and back again every hour, he or she must become startlingly aware of how artificial the boundaries are that we've created to separate and define ourselves.

So here we are, with our eyes fixed on the stars, in a dilemma that becomes increasingly terrifying. Will we have the wisdom and courage to accept the moral reality within each of us? Or will we defer to the illusory corporeal reality in the false belief that it contains all the answers we need?

Our answers to these questions may have a profound effect on the outcome of this greatest of all experiments—life.

And so, my dear reader, read on.

Rav Berg
New York, 1987
The Power of *Alef Bet*

INTRODUCTION

ASK THEE A SIGN OF THE
LORD; ASK IT EITHER
IN THE DEPTH BELOW OR
IN THE HEIGHT ABOVE.

—ISAIAH 7:11

Early generations of the children of Israel commonly knew what today only a handful of people understand—the profound internal power of the twenty-two letters of the Hebrew *Alef Bet*. Since the time of Abraham, the Hebrews demonstrated an uncanny awareness of rudimentary physics. Although they didn't express it in the language of modern mathematics and science, early kabbalists applied their understanding of the universal energy fields of positive, negative, and neutral to the circuitry and workings of the hearts and minds of humankind. The negative side of the energy field was given the name Desire to Receive for Oneself Alone; the positive side was called the Desire to Receive for the Sake of Imparting.

Energy fields have intelligence, in much the same way that we understand ourselves to possess intelligence. Everything that exists in all of Creation—from stones to stars, from plants to animals to man—every speck of Creation is composed not merely of energy, but of infinite energy-intelligence. When we harness this powerful resource, we energize the principle of "Love thy neighbor," creating balance and health in our lives and in the lives of those who choose to acknowledge this energy.

Kabbalah teaches that the idea of the wheel came before the invention of the wheel. It is thoughts and ideas that enable us to create the physical world, as well as influence what occurs in the cosmos. We understand that the moon affects the tides. We acknowledge that supernovas, black holes, and other phenomena in outer space inevitably affect weather and other physical conditions here on Earth. But can we comprehend the ancient kabbalistic teaching that our behavior can override extraterrestrial influences and hold sway over intergalactic events?

The destruction by the Romans of the Second Temple in Jerusalem in the year 70 CE nearly put an end to the possibilities opened up by Kabbalah. But this vital, ancient wisdom could never be extinguished. During the long centuries since that time, the Light of the wisdom of Kabbalah has flickered. It is written in the *Zohar (Book of Splendor)* that Kabbalah would have to await the coming of the Age of Aquarius to make its reappearance as a tool that we can use to draw down the Lightforce of the Creator to humankind, as they currently wander in cosmic Darkness.

This time has come! The *Zohar* is a book of power—the power to activate the letters of the *Alef Bet* and to make them do our bidding!

The Hebrew letters—twenty-two distinct and astonishingly powerful energy-intelligences—are animated by a spiritual force more immense than the energy contained within the atom. But the *Alef Bet* is of no practical use if we do not understand how to connect ourselves to its inherent energy. Kabbalah is a technology that gives us access to this life-affirming, all-powerful system.

The truth is that we've already been penetrated by the cosmic vitality of the *Alef Bet*. Kabbalists have been saying this for centuries, but the scientific community didn't catch on until 1926. This was the year German physicist Werner Heisenberg developed his now-famous Uncertainty Principle, which states that it is impossible to precisely measure both the position and the momentum of a particle at the same time. Heisenberg found that the more precisely you measure a particle's position, the less precisely you can measure its momentum, and vice versa. One interesting ramification of Heisenberg's discovery is that an observer can influence—indeed, must influence—a phenomenon simply by the act of observing it.

This conclusion flies in the face of Newtonian science, which assumes the scientist is detached from the experiments he conducts and observes. Heisenberg's Uncertainty Principle instead affirms the ancient kabbalistic tenet that the observer and the observed are inseparably, irretrievably bound up with one another. Everything is part of a unified whole, as the Bible declares in the *Shema*: "Hear, Israel, the Lord our God, the Lord is One."

Humankind and cosmos are not the separate, distinct entities that Newtonian science would have us believe. Albert Einstein, with his Theory of Relativity, was the first physicist to take us beyond the bounds of Newtonian science, and he spent the next fifty years trying to go even further, seeking a Unified Field Theory that would unite all known energies in a single complete description. Einstein died without achieving his goal, but in the meantime, Newtonian science, detached from the moral consequences of its findings, has brought us to the edge of nuclear disaster.

The issue of nuclear war is not a matter solely for world leaders and diplomats to somehow resolve. World peace is as much a matter of the heart as it is of politics. We read our newspapers over our morning coffee and lament the state of the world, little knowing that we all share the burden of nuclear responsibility, never dreaming that the perennial drama of war and peace is acted out every day by the manner in which we conduct our lives. Peace on Earth will never be realized if we cannot achieve peace within ourselves. If our households, offices, and factories remain battlegrounds of human misery, world peace cannot be achieved because world peace can never be more than the aggregate of the internal, personal peace experienced by each of us. To survive as a species, we must accept what has been known since Abraham smashed his father's idols: God is ONE, and everything that exists has inherited its portion of the Supreme Intelligence.

The answers to life's most intractable mysteries are bound up with the way we think. Oliver Wendell Holmes said, "Man's mind, once stretched by a new idea, never regains its original dimensions." We have now encountered these new ideas, and thus, I believe the time has come when we must begin to contemplate a flight into space—not to outer space, but to inner space, where all that exists truly originated: the inner space of *thought*. For it is here—in the arena of human notions and emotions—that the Lightforce will either conquer or be defeated by the Dark Lord who has manned his battle stations in readiness for combat.

Many embrace this vision; others perceive it as fantasy. But those who believe that explorations of mind and thought are futile or impossible are ignoring both ancient wisdom and scientific evidence that supports two key metaphysical truths: the realm of thought is the true reality, and man and the cosmos are one and the same.

Though earth and man were gone,
And suns and universes ceased to be,
And Thou wert left alone,
Every existence would exist in Thee.
—Emily Bronte

Since time immemorial, people have searched for possible relationships between the arrangement and motion of celestial bodies and events that occur on Earth. Traditionally, the search for the cosmic connection was the domain of the astrologer because the scientific community either maligned or rejected outright the mystique surrounding the heavens. Following the introduction of quantum theory, however, these rigid, Newtonian attitudes changed dramatically—so much so that today almost anything, from parallel universes to time running backward, is believed to be possible.

Much of what Kabbalah says about the structure of matter resembles the worldview of the quantum physicist. Yet the quantum view falls far short of the kabbalistic account of the laws that govern our universe. The quantum physicist admits that there is no physical basis for matter. Yet the concept that consciousness might transcend the solidity of a table to interact with waves or particles as if they were components of our imagination is still too far removed from normal experience to be readily embraced by even the most avant-garde physicist, much less by the man on the street. To our rational Western way of thinking, the transcendent properties of physical matter would have to be considered a miracle. Proof that consciousness is capable of creating and influencing matter would give credence to the kabbalistic teaching that the metaphysical world has dominion over the physical, corporeal world. From a kabbalistic point of view, matter is the condensed energy of consciousness itself.

Does the world continue on as it was even if we do not perceive it? When we wake up in the morning, does the same world exist as the one we fell asleep in yesterday? The Copenhagen interpretation of quantum physics, as presented by Niels Bohr in 1927, laid to rest the classical idea of an objectively existing universe. No longer would scientists naively accept the idea that the world had an objective existence independent of our observation of it. The Copenhagen interpretation maintains that at the subatomic level, the world we see depends on how we observe it—and, more importantly, on what we choose to see. In other words, our observations can change what we're observing. The participatory nature of the quantum world implies that the traditional scientists' view of physical phenomena as external objects must be replaced by a new observer-centric account of reality.

The teachings of Rav Isaac Luria (the Ari, 1534–1572), the founder of Lurianic Kabbalah, presaged this radical new role of consciousness in physics. Indeed, the Ari went even further when he stated that the terms "observer" and "participator" should be replaced by the term "determinator."[1] In a similar vein, physicist Jack Sarfatti wrote: "An idea of the utmost significance for the development of psycho-energetic systems is that the structure of matter may not be independent of consciousness."[2]

In his *Gate of Divine Inspiration*, the Ari describes what must occur when consciousness affects matter:

When a person does a good deed, he makes manifest and acquires a personal positive intelligent life force. All essence within our universe has been structured by the actions of man. For even the sound that emanates from the striking of a stone by the rod is not in vain. It maintains its rightful place in the cosmos. From even man's word of mouth are created angelic, metaphysical life forces. These very same forces become chariots integrated with the whole of the cosmos. They then connect with the souls of the righteous of the past. Through this interconnectedness, these life-form energy-intelligences then serve as providers of cosmic intelligences. They assist the creator of these forces (man), which have become chariots for cosmic intelligences.[3]

The conscious mind is a micro-system of the cosmic macro-system, which consists entirely of thought-information and life-form intelligences—an all-embracing unified reality to which each of us can connect. Physicist John Wheeler pointed toward this reality when he said, "The beauty in the laws of physics is their fantastic simplicity. What is the ultimate mathematical machinery behind it all? That must surely be the most beautiful of all."[4] Einstein voiced a similar sentiment when he remarked, "All of these endeavors are based on the belief that existence should have a completely harmonious structure. Today we have fewer grounds than ever before for allowing ourselves to be forced away from this wonderful belief."[5] Thus are our greatest scientists inspired by the beauty of the natural world they seek to understand.

Kabbalah offers a definite path to follow in our search for answers to longstanding scientific questions. Unlike science, which is concerned solely with the *how* of things, Kabbalah is concerned with the *why*. It is my deep conviction that we can begin to comprehend the meaning of the fundamental truths and laws of our universe only once we discover the answers to why they exist. There is only one way we can gain these insights: through metaphysical, cosmic connections. And it is unconscious humankind, when connecting with cosmic reality, that acts as the channel for the revelation of reality.

Despite the vast amount of scientific research directed at the cosmos, science is still light-years away from elucidating the human link between Heaven and Earth. On the other hand, the kabbalistic worldview already provides another approach to illuminating the cosmic Darkness. The central life-force entities of the universe were created through the energy-intelligences inherent in the letters of the Hebrew *Alef Bet*. The elements of fire, water, air, and earth came into being through the manifestation of these letter-energies. Through these twenty-two living, breathing entities, we can at last glimpse the Light at the end of the cosmic tunnel.

Abraham the Patriarch, the world's first known astrologer, was initiated into the mysteries of cosmological speculation. He was taught the deeper meanings hidden within the permutations and combinations of the Hebrew letters and the numbers they represent. Various commentators throughout the ages have expounded upon the science of the combination of letters—a way to control the internal forces of our universe with the aid of letters and their configurations. For the kabbalist, the Divine language, formed by the *Alef Bet*, is the substance of reality.

The letters of this spiritual vocabulary are the elements of the profoundest, most fundamental level of intelligence and understanding. Contemplation of these letters leads to an awareness of the unity and mutual interrelationship of all phenomena, and therefore to new states of consciousness. Thus, the *Alef Bet* provides direct experience of all that exists as an interlocking manifestation of a single, all-pervasive cosmic Unity.

Abraham declared that the Hebrew *Alef Bet* is the cosmic bond. With the revival of this ancient wisdom, the search for the unifying force of our cosmos may at last come to an end.

MYSTICAL HISTORY OF THE HEBREW LETTERS

FROM THE *ZOHAR*
PROLOGUE 6:23-36

The letters by Rav Hamnuna Saba

Individual energy forces, which express themselves as Hebrew letters, came before the Creator requesting that they be the instruments by which the world is created. The Creator eventually agrees to utilize the letter *Bet* בָ, as this particular letter begins the Hebrew word *Berachah* [Blessing]. The *Zohar* then depicts the unique attributes of each of the twenty-two letters and the spiritual energy they emit. All these forces and their power of blessing are transferred to us when we visually scan the Aramaic text and learn the lessons it holds.

23. כַּד בְּעָא לְמִבְרֵי עָלְמָא, אָתוּ כָּל אַתְוָון קַמֵּיהּ מִסּוֹפָא אֲרֵישַׁיְיהוּ. שָׁרֵיאַת אָת ת לְמֵיעַל בְּרֵישָׁא, אָמְרָה, רִבּוֹן עָלְמִין: נִיחָא קַמָּךְ לְמִבְרֵי בִּי עָלְמָא, דְּאֲנָא חוֹתָמָא דְגוּשְׁפַּנְקָא דִילָךְ, אֱמֶת, וְאַתְּ אִתְקְרֵיאַת אֱמֶת, יָאוֹת לְמַלְכָּא לְמִשְׁרֵי בְּאוֹת אֱמֶת, וּלְמִבְרֵי בִּי עָלְמָא. אָמַר לָהּ קֻדְשָׁא בְּרִיךְ הוּא יָאוֹת אַנְתְּ וְזַכָּאָה אַנְתְּ, אֶלָּא לֵית אַנְתְּ כְּדַאי לְמִבְרֵי בָּךְ עָלְמָא. הוֹאִיל וְאַנְתְּ זַמִּינָא לְמֶהֱוֵי רָשִׁים עַל מִצְחִין דְּגוּבְרִין מְהֵימְנִין, דְּקַיְּימוּ אוֹרַיְיתָא מָא' וְעַד תנ', וּבְרָשִׁימוּ דִילָךְ יְמוּתוּן. וְעוֹד, דְּאַנְתְּ חוֹתָמָא דְמָוֶת, הוֹאִיל וְאַנְתְּ כָּךְ, לֵית אַנְתְּ כְּדַאי לְמִבְרֵי בָּךְ עָלְמָא. מִיָּד נָפְקַת.

22. Beresheet: Rav Hamnuna Saba, said that we have found IN THE WORDS: "In the beginning, God created the... (HEB. *BERESHEET* BARA *ELOHIM*)." (Genesis 1:1) THAT THE ORDER OF the letters IN THIS PHRASE IS ARRANGED in reverse. First, the Hebrew letter *Bet* is immediately followed by another *Bet*, that is, *Beresheet bara*. Subsequently, IT IS WRITTEN first with an *Alef* and then another *Alef*, referring in Hebrew to *Elohim*. AND HE EXPLAINS THAT when the Holy One, blessed be He, was about to create the world, all of the letters were STILL hidden. For two thousand years before the creation of the world, the Holy One, blessed be He, watched the letters and amused Himself with them.

ת

(A) The letter *Tav*

23. כַּד בָּעָא לְמִבְרֵי עָלְמָא, אָתוּ כָּל אַתְוָון קַמֵּיה מִסּוֹפָא אֲרֵישַׁיְיהוּ. שָׁרֵיאַת אָת **ת** לְמֵיעַל בְּרֵישָׁא, אָמְרָה, רִבּוֹן עָלְמִין: נִיחָא קַמָּךְ לְמִבְרֵי בִּי עָלְמָא, דַּאֲנָא חוֹתָמָא דְגוּשְׁפַּנְקָא דִילָךְ, אֱמֶת, וְאַתְּ אִתְקְרֵיאַת אֱמֶת, יָאוֹת לְמַלְכָּא לְמִשְׁרֵי בְּאוֹת אֱמֶת, וּלְמִבְרֵי בִּי עָלְמָא. אָמַר לָהּ קֻדְשָׁא בְּרִיךְ הוּא יָאוֹת אַנְתְּ וְזַכָּאָה אַנְתְּ, אֶלָּא לֵית אַנְתְּ כְּדַאי לְמִבְרֵי בָּךְ עָלְמָא. הוֹאִיל וְאַנְתְּ זַמִּינָא לְמֶהֱוֵי רָשִׁים עַל מִצְחִין דְּגוּבְרִין מְהֵימָנִין, דְּקַיְימוּ אוֹרַיְיתָא מֵ**א'** וְעַד **ת'**, וּבְרְשִׁימוּ דִּילָךְ יְמוּתוּן. וְעוֹד, דְּאַנְתְּ חוֹתָמָא דְמָוֶת, הוֹאִיל וְאַנְתְּ כָּךְ, לֵית אַנְתְּ כְּדַאי לְמִבְרֵי בָּךְ עָלְמָא. מִיָּד נָפְקַת.

23. When He desired to create the world, all the letters came before Him in sequence from last to first. The letter *Tav* saw itself fit to come forth first. It said, Master of the World, may it please You to create the world with me because I am the seal of Your ring, which is *emet* (truth), MEANING THAT I AM THE LAST LETTER IN THE WORD *EMET*. And as You are called by this name of truth, so it would be most appropriate for the King to begin with the letter *Tav*, and create the world by me. The Holy One, blessed be He, said to it, You are worthy and deserving, but you are not suitable for the world to be created by you. You are destined to serve as a mark on the foreheads of the faithful, who have kept the Law of the Torah from *Alef* to *Tav*. When you appear, they shall die. Not only that, but you are the seal of the word Death, MEANING THAT *TAV* IS THE LAST LETTER IN THE WORD *MAVET* (DEATH). And because of this, you are not suitable for Me to create the world with you. The letter *Tav* then immediately left.

שׁ

(B) The letter *Shin*

24. עָאלַת אָת **שׁ** קַמֵּיהּ, אָמְרָה קַמֵּיהּ: רִבּוֹן עָלְמִין, נִיחָא קַמָּךְ
לְמִבְרֵי בִּי עָלְמָא, דְּבִי אִתְקְרֵי שְׁמָךְ שַׁדַּ"י, וְיָאוֹת לְמִבְרֵי עָלְמָא
בִּשְׁמָא קַדִּישָׁא. אָמַר לָהּ: יָאוֹת אַנְתְּ וְטַב אַנְתְּ וּקְשׁוֹט אַנְתְּ, אֲבָל
הוֹאִיל וְאַתְוָון דְּזִיּוּפָא נַטְלִין לָךְ לְמֶהֱוֵי עִמְּהוֹן לָא בָּעֵינָא לְמִבְרֵי
בָּךְ עָלְמָא, דִּבְגִין דְּלָא יִתְקַיַּים שִׁקְרָא אֶלָּא אִי יִטְלוּן לָךְ **ק, ר.**

24. The letter *Shin* then entered and stood before Him. It said, Master of the Universe, may it please You to create the world with me, as Your own name *Shaddai* is with me. And it would be most proper to create the world by a holy name. He replied, You are worthy, you are good, and you are truthful. But because you are included among the letters that form the word *sheker* (falsehood), I do not wish to create the world with you. *Sheker* would not have prevailed had you not been attached to the letters RESH and KUF.

(C) The letter *Kuf* and the letter *Resh*

25. מִכָּאן, מָאן דְּבָעֵי לְמֵימַר שִׁקְרָא יִטוֹל יְסוֹדָא דִקְשׁוֹט
בְּקַדְמֵיתָא, וּלְבָתַר יוֹקִים לֵיהּ שִׁקְרָא, דְּהָא אָת שׁ אָת קְשׁוֹט אִיהוּ,
אָת קְשׁוֹט דַּאֲבָהָתָן דְּאִתְיַיחֲדוּ בֵּהּ ק, ר אַתְוָון דְּאִתְחֲזִיאוּ עַל
סִטְרָא בִּישָׁא אִינּוּן, וּבְגִין לְאִתְקַיְּימָא נָטְלֵי אָת שׁ בְּגַוַּויְיהוּ הֲוֵי
קֶשֶׁר. כֵּיוָן דְּחָמָאת הָכֵי נָפְקַת מִקַּמֵּיהּ.

25. From this WE LEARN THAT whoever wants to tell a lie, should add that lie to a base that is truthful. Because the letter *Shin* is a letter (namely a sign) of truth, meaning a letter of Truth by which the Patriarchs reached unison. NOTE THAT THE THREE LINES IN THE LETTER SHIN REFER TO THE THREE PATRIARCHS, WHO ARE CALLED *CHESED*, *GEVURAH*, AND *TIFERET*. And the letters *KUF* AND *RESH* appear on the evil side, BECAUSE THE OTHER SIDE IS COLD (HEB. *KAR—KUF*, *RESH*), WITHOUT ANY OF THE WARMTH THAT GIVES LIFE. THE OTHER SIDE DRAWS ITS SUSTENANCE FROM *MALCHUT* WHEN SHE IS A FROZEN OCEAN. In order for them to continue to exist, they pulled the letter *Shin* to them, creating the combination *Kesher*, WHICH MEANS 'STRENGTHENING' AND 'SURVIVING.' When the *Shin* realized this, it left.

(D) The letter *Tzadi*

‏26. עָאלַת אָת צ אָמְרָה קַמֵיהּ: רִבּוֹן עָלְמָא, נִיחָא קַמָךְ לְמִבְרֵי בִּי‏
‏עָלְמָא, דְּאֲנָא, בִּי חֲתִימִין צַדִּיקִים, וְאַנְתְּ, דְּאִתְקְרִיאַת צַדִּיק, בִּי‏
‏רְשִׁים, דִּכְתִיב כִּי צַדִּיק ה' צְדָקוֹת אָהֵב, וּבִי יָאוֹת לְמִבְרֵי עָלְמָא.‏
‏אָמַר לַהּ: צַדִּי, צַדִּי אַנְתְּ, וְצַדִּיק אַנְתְּ, אֲבָל אַנְתְּ צָרִיךְ לְמֶהֱוֵי‏
‏טְמִירָא, לֵית אַנְתְּ צָרִיךְ לְאִתְגַּלְיָא כָּל כָּךְ, בְּגִין דְּלָא לְמֵיהַב פִּתְחוֹן‏
‏פֶּה לְעָלְמָא. מ"ט צ אִיהִי, אַתְיָא י דִּשְׁמָא דִּבְרִית קַדִּישָׁא וְרָכִיב‏
‏עֲלַהּ וְאִתְאֲחַד בַּהֲדָהּ. וְרָזָא דָא, כַּד בְּרָא קַדְשָׁא בְּרִיךְ הוּא לְאָדָם‏
‏הָרִאשׁוֹן דּוּ פַּרְצוּפִין בְּרָאוֹ. וּבְגִין כָּךְ אַנְפּוֹי דִיו"ד מְהַדַּר לַאֲחוֹרָא‏
‏כְּגַוְונָא דָא צ, וְלָא אִתְהַדְרוּ אַנְפִּין בְּאַנְפִּין כְּגַוְונָא דָא צ, אִסְתַּכַּל‏
‏לְעֵילָא כְּגַוְונָא דָא צ אִסְתַּכְּלַת לְתַתָּא כְּגַוְונָא דָא צ אָמַר לַהּ‏
‏קַדְשָׁא בְּרִיךְ הוּא: תּוּ, דַּאֲנָא זַמִּין לְנַסְרָא לָךְ, וּלְמֶעְבַּד לָךְ אַפִּין‏
‏בְּאַפִּין, אֲבָל בְּאַתְרָא אָחֳרָא תִּסְתַּלָק. נָפְקַת מִקַּמֵיהּ וַאֲזָלַת.‏

26. The letter *Tzadi* then entered, stood before Him, and said, Master of the Universe, may it please You to create the world with me, since the righteous (Heb. *tzadikim*) are 'signed' by my name. And You, who are called righteous (Heb. *Tzadi*), are also written by my name, as it is written, "For God is righteous, He loves righteousness." (Psalms 11:7) Therefore, it would be suitable to create the world with me! He replied, *Tzadi*, *Tzadi* you are truly righteous, but you should remain concealed and not be revealed too much; AS WOULD BE DONE IF THE WORLD WERE CREATED BY YOU, so that humans will not have an excuse for their sins. And what is the reason FOR IT REMAINING CONCEALED? IT IS BECAUSE it used to be *Nun*, and the letter of the Name, which is the Holy Covenant, came and mounted on the letter Nun, THEREBY CREATING THE LETTER *TZADI*. The secret meaning of this matter is that when the Holy One, blessed be He, created Adam, WHO IS THE SECRET OF ZEIR ANPIN, He created him with two faces, THAT IS A MALE AND A FEMALE ATTACHED AT THE BACK. For that reason the *Yud*'s face was turned away FROM THE *NUN*, THE *YUD* FACING ONE WAY, THE NUN ANOTHER, not turning to face each other. The *Yud* faced upwards AND THE NUN faced downwards. The Holy One, blessed be He, said to it: I will sew you—THAT IS, SEPARATE THE ATTACHMENT OF BACK-TO-BACK WITHIN YOU and form you into a face-to-face union. But it will come to be this way in another place, NOT IMMEDIATELY WITH THE CREATION OF THE WORLD. SINCE BEING JOINED BACK-TO-BACK IS AN INDICATION THAT ITS ILLUMINATION IS HIDDEN, IT IS NOT SUITABLE FOR THE CREATION OF THE WORLD. THE LETTER *TZADI* left His presence, and went on its way.

עַ פ

(E) The letter *Pei* and the letter *Ayin*

‎27. עָאלַת פ אָת אָמְרָה קַמֵּיהּ: רִבּוֹן עָלְמִין, נִיחָא קַמָּךְ לְמִבְרֵי בִּי
עָלְמָא, דְּהָא פּוּרְקָנָא דְּאַנְתְּ זַמִּין לְמֶעֱבַד בְּעָלְמָא, בִּי רְשִׁים, וְדָא
הוּא פְדוּת, וּבִי יָאוֹת לְמִבְרֵי עָלְמָא. אָמַר לָהּ: יָאוֹת אַנְתְּ, אֲבָל בָּךְ
אִתְרְשִׁים פֶּשַׁע בִּטְמִירוּ, כְּגַוְונָא דְחִיוְיָא דְּמָחֵי, וְאָעֵיל רֵישֵׁיהּ בֵּין
גּוּפֵיהּ, הָכִי, מַאן דְּחָב, כָּפִיף רֵישֵׁיהּ וְאַפִּיק יְדוֹי. וְכֵן עַ עָוֹן, אַף
עַל גַּב דְּאָמְרָה, דְּאִית בִּי עֲנָוָה, אָמַר לָהּ קֻדְשָׁא בְּרִיךְ הוּא לָא
אִבְרֵי בָךְ עָלְמָא. נָפְקַת מִקַּמֵּיהּ.

27. The letter *Pei* was the next to enter. It stood before Him and said, Master of the Universe, may it please You to create the world with me, because the Redemption that You shall bring to the world is described by me, *Pedut* (Redemption). THIS MEANS THAT SALVATION IS TO BE REDEEMED FROM OUR ENEMIES, AND THIS WORD IN HEBREW STARTS WITH THE LETTER *PEI*. THIS IS WHY the world should be created with me. He replied to it, You are indeed praiseworthy, but a secret crime (Heb. *pesha*) is inscribed in you, like the snake that strikes and then brings its head back within the coils of its body. Because whoever sins bends his head, MEANING THAT HE HIDES HIMSELF FROM THE 'OBSERVING EYE,' then stretches his hands out TO SIN. THIS REFERS TO THE SHAPE OF THE LETTER *PEI* THAT HAS A HEAD BENT DOWN INTO ITS BODY. AND SO IT WAS SIMILARLY SAID OF THE LETTER *AYIN*, WHICH DESCRIBES THE TERM *avon* (iniquity). Although it claimed, I have humility (Heb. *anavah*) in me, the Holy One, blessed be He, replied to it saying, I shall not create the world by you. *Ayin* then took its leave!

ס

(F) The letter *Samech*

28. עָאלַת אָת ס אָמְרָה קַמֵּיה: רִבּוֹן עָלְמִין, נִיחָא קַמָּךְ לְמִבְרֵי בִּי
עָלְמָא, דְּאִית בִּי סְמִיכָא לְנָפְלִין, דִּכְתִיב סוֹמֵךְ ה' לְכָל הַנּוֹפְלִים.
אָמַר לָהּ: עַל דָּא אַנְתְּ צָרִיךְ לְאַתְרָךְ, וְלָא תְזוּז מִנֵּיה, אִי אַתְּ נָפִיק
מֵאַתְרָךְ, מַה תְּהֵא עֲלַיְיהוּ דְּאִינּוּן נְפִילִין, הוֹאִיל וְאִינּוּן סְמִיכִין
עֲלָךְ. מִיָּד נָפְקַת מִקַּמֵּיה.

28. The letter *Samech* (support) entered, stood before Him and said, Master of the Universe, may it please You to create the world with me, because I am able to support those who fall. As it is written, "God upholds (Heb. *somech*) all that fall." (Psalms 145:14) He said to it, This is exactly why you should stay in your place and not move from it. If you leave your place IN THE WORD SOMECH, what will then happen to all those who fell and are being supported by you? The letter *Samech* then left immediately!

(G) The letter *Nun*

29. עָאלַת אָת **נ** אָמְרָה קַמֵּיה רִבּוֹן עָלְמָא, נִיחָא קַמָּךְ לְמִבְרֵי בִּי
עָלְמָא, דְּבִי כְּתִיב נוֹרָא תְהִלּוֹת, וּתְהִלָּה דְּצַדִּיקִים נָאוָה תְהִלָּה.
אָמַר לָהּ: נוּ"ן, תּוּב לְאַתְרָךְ דְּהָא בְּגִינָךְ תָּבַת סָמַ"ךְ לְאַתְרָהּ, וַהֲוֵי
סָמִיךְ עֲלָהּ. מִיָּד תָּבַת לְאַתְרָהּ וְנָפְקַת מִקַּמֵּיה.

29. The letter *Nun* entered and stood before Him saying, Master of the Universe, may it please You to create the world with me, because the phrase, "fearful in praises (Heb. *nora tehilot*)," (Exodus 15:11) starts with me. AND ALSO, in the praising of the righteous, IT IS WRITTEN, "Praise (Heb. *nava*) is comely." (Psalms 147:1) He said, *Nun,* go back to your place. It is because of you that the letter *Samech* returned to its place. And you should depend on it for support. THIS MEANS THAT THE LETTER *NUN* IS INSCRIBED AT THE BEGINNING OF THE WORD FALL (HEB. *NEFILAH*) AND THE LETTER *SAMECH*, WHICH IS THE SECRET OF "GOD UPHOLDS ALL THAT FALL," RETURNED TO ITS PLACE TO SUPPORT THOSE WHO FALL, AS EXPLAINED IN THE PREVIOUS PARAGRAPH. It immediately left His presence and returned to its place.

מ ל

(H) The letter *Mem* and the letter *Lamed*

30. עָאלַת אָת **מ** אָמְרָה קַמֵּיהּ: רִבּוֹן עָלְמָא, נִיחָא קַמָּךְ לְמִבְרֵי בִּי עָלְמָא, דְּבִי אִתְקְרִיאַת מֶלֶךְ. אָמַר לָהּ: הָכֵי הוּא וַדַּאי, אֲבָל לָא אִבְרֵי בָּךְ עָלְמָא, בְּגִין דְּעָלְמָא אִצְטְרִיךְ לְמֶלֶךְ, תּוּב לְאַתְרָךְ, אַנְתְּ וְ ל ו ך, דְּהָא לָא יָאוֹת לְעָלְמָא לְמֵיקַם בְּלָא מֶלֶךְ.

30. The letter *Mem* entered and said to Him, Master of the Universe, may it please You to create the world with me, because with me you are called *Melech* (King). He told it, Yes, it is indeed so, but I shall not create the world with you, because the world needs a king! Go back to your place, you and the letters *Lamed* and *Kaf*, as it is not proper for the world to be without a king.

(I) The letter *Kaf*

‏31. בְּהַהִיא שַׁעֲתָא, נָחֲתָא מִן קֳדָמוֹהִי אָת כ מֵעַל כּוּרְסְיֵהּ יְקָרֵיהּ,
אִזְדַּעְזָעַת וְאָמְרָה קַמֵּיהּ: רִבּוֹן עָלְמָא, נִיחָא קַמָּךְ לְמִבְרֵי בִּי
עָלְמָא, דַּאֲנָא כְּבוֹדָךְ. וְכַד נָחֲתַת כ מֵעַל כּוּרְסְיֵהּ יְקָרֵיהּ, אִזְדַּעְזָעוּ
מָאתָן אֶלֶף עָלְמִין וְאִזְדַּעְזַע כֻּרְסְיָיא, וְכֻלְּהוּ עָלְמִין אִזְדַּעְזָעוּ
לְמִנְפָּל. אָמַר לָהּ קֻדְשָׁא בְּרִיךְ הוּא: כ"ף, כ"ף, מָה אַתְּ עָבֵיד
הָכָא, דְּלָא אִבְרֵי בָּךְ עָלְמָא, תּוּב לְאַתְרָךְ, דְּהָא בָּךְ כְּלָיָה, כָּלָה
וְנֶחֱרָצָה אִשְׁתְּמַע, תּוּב לְכֻרְסָיָךְ וֶהֱוֵי תַמָּן. בְּהַהִיא שַׁעֲתָא נָפְקַת
מִקַּמֵּיהּ וְתָבַת לְדוּכְתָּהּ.

31. At that hour, the letter *Kaf* descended from His Throne (Heb. *kise*) of
Glory. Shaking and trembling, it stood before Him and said, Master of the
Universe, may it please You to create the world with me, because I am Your
Glory (Heb. *kavod*). When the letter *Kaf* descended from the Throne of His
Glory, 200,000 worlds were shaken and the Throne trembled. And all the
worlds were about to collapse. The Holy One, blessed be He, told it, *Kaf, Kaf,*
what are you doing here?! I shall indeed not create the world with you. Go
back to your place, because the Hebrew word *klayah* (destruction) starts with
you. AND BECAUSE OF YOU "...total destruction is determined and decreed."
(Isaiah 10:23) So, return to your Throne and stay there. At that same
moment, it took leave and returned to its place.

(J) The letter *Yud*

‫32. עָאלַת אָת י אָמְרָה קַמֵּיה: רִבּוֹן עָלְמָא, נִיחָא קַמָּךְ לְמִבְרֵי בִּי‬
‫עָלְמָא, דַּאֲנָא שֵׁירוּתָא דִּשְׁמָא קַדִּישָׁא, וְיָאוֹת לָךְ לְמִבְרֵי בִּי‬
‫עָלְמָא. אָמַר לָהּ: דִּי לָךְ דְּאַנְתְּ חָקִיק בִּי, וְאַנְתְּ רָשִׁים בִּי, וְכָל‬
‫רְעוּתָא דִּילִי בָּךְ, סָלֵיק, לֵית אַנְתְּ יָאוֹת לְאִתְעַקְּרָא מִן שְׁמִי.‬

32. The letter *Yud* entered, stood before Him, and said, Master of the Universe, may it please You to create the world with me, because I am the first letter of the Holy Name. Thus, it should be proper for You to create the world with me. He replied, It should suffice you to be engraved upon My Name and appear in Me. You embrace all My desires. Rise up, it would not be proper for you to be removed from My Name!.

(K) The letter *Tet* and the letter *Chet*

‏33. עָאלַת אָת **ט** אָמְרָה קַמֵיה: רִבּוֹן עָלְמָא, נִיחָא קַמָּךְ לְמִבְרֵי בִּי
עָלְמָא, דְּאַנְתְּ, בִּי אִתְקְרִיאַת טוֹב וְיָשָׁר. אָמַר לָה: לָא אִבְרֵי בָּךְ
עָלְמָא, דְּהָא טוּבָךְ סָתִים בְּגַוָּוךְ וְצָפוּן בְּגַוָּוךְ, הה"ד מָה רַב טוּבְךָ
אֲשֶׁר צָפַנְתָּ לִּירֵאֶךָ, הוֹאִיל וְגָנִיז בְּגַוָּוךְ, לֵית בֵּיהּ חוּלָקָא לְעָלְמָא
דָא, דַּאֲנָא בָּעֵי לְמִבְרֵי, אֶלָּא בְּעָלְמָא דְאָתֵי. וְתוּ, דְּעַל דְּטוּבָךְ גָּנִיז
בְּגַוָּוךְ, יִטְבְּעוּן תַּרְעֵי דְהֵיכָלָא. הה"ד טָבְעוּ בָאָרֶץ שְׁעָרֶיהָ. וְתוּ דְּ
וֹ לְקִבְלָךְ, וְכַד תִּתְחַבְּרוּן כַּחֲדָא, הָא ח'ט, וְעַל דָּא אַתְוָון אִלֵּין
לָא רְשִׁימִין בְּשִׁבְטִין קַדִּישִׁין, מִיָּד נָפְקַת מִקַּמֵיהּ.

33. The letter *Tet* entered, stood before Him and said, Master of the Universe may it please You to create the world with me, as by me You are called good (Heb. *tov*) and upright. He replied: I will not create the world with you, because your goodness is concealed within you. Therefore it is written, "How abundant is Your goodness which You have concealed for them that fear You." (Psalms 31:20) So because your goodness is concealed within you, it cannot take any part in this world that I want to create. It only applies to the World to Come. Furthermore, because your goodness is concealed and treasured within yourself, the gates of the Temple shall be 'sunk.' As it is written, "Her gates are sunk into the ground." (Lamentations 2:9) And to add to all this, the letter *Chet* stands before you, together you become a sin (Heb. *Chet—Chet, Tet, Alef*). This is why these two letters do not appear in the names of the twelve tribes. Tet immediately then took its leave and went away from Him.

(L) The letter *Zayin*

‏34. עָאלַת אָת ז אָמְרָה לֵיהּ: רִבּוֹן עָלְמָא, נִיחָא קַמָּךְ לְמִבְרֵי בִּי
עָלְמָא, דְּבִי נָטְרִין בְּנָיךְ שַׁבָּת, דִּכְתִיב זָכוֹר אֶת יוֹם הַשַּׁבָּת לְקַדְּשׁוֹ.
אָמַר לָהּ: לָא אַבְרֵי בָּךְ עָלְמָא, דְּאַנְתְּ אִית בָּךְ קְרָבָא, וְחַרְבָּא
דְּשַׁנְנָא, וְרוֹמְחָא דִקְרָבָא, כְּגַוְונָא דְנוּן, מִיָּד נָפְקַת מִקַּמֵּיהּ.

34. The letter *Zayin* entered and said to Him, Master of the Universe, may it please You to create the world with me, because with my help, Your children shall keep the Shabbat, as it is written, "Remember (Heb. *zachor*) the *Shabbat* day, to keep it holy." (Exodus 20:8) He replied, I will not create the world with you, because you represent war, that is, a sharp pointed sword and a spear with which people make war. AND THEY ARE CALLED WEAPON, WHICH IN HEBREW IS PRONOUNCED *ZAYIN!* And you are like the letter *Nun*, WHICH THE WORLD WAS NOT CREATED BY, BECAUSE IT IS AT THE BEGINNING OF THE WORD *NEFILAH* (FALLING). It immediately left His presence.

(M) The letter *Vav* and the letter *Hei*

‎35. עָאלַת אָת וֹ אָמְרָה קַמֵּיהּ: רִבּוֹן עָלְמָא, נִיחָא קַמָּךְ לְמִבְרֵי בִּי
‎עָלְמָא, דַּאֲנָא אָת מִשְּׁמָךְ. אֲמַר לָהּ: וָאו, אַנְתְּ וּ הֹ, דִּי לְכוֹן דְּאַתּוּן
‎אַתְוָון דִּשְׁמִי, דְּאַתּוּן בְּרָזָא דִשְׁמִי, וַחֲקִיקִין וּגְלִיפִין בִּשְׁמִי, וְלָא
‎אִבְרֵי בְּכוֹ עָלְמָא.

35. The letter *Vav* entered, and pleaded before Him, Master of the Universe, may it please You to create the world with me because I am one of the letters of Your Name *YUD, HEI, VAV,* AND *HEI*. He replied, Vav, you and the letter *Hei* should both be satisfied with being written in My Name. Because you appear in My Name and are engraved in It, I shall therefore not create the world with you.

(N) The letter *Dalet* and the letter *Gimel*

‏36. עָאלַת אָת דֹ וְאָת גֹ אָמְרוּ אוֹף הָכֵי, אָמַר אוֹף לוֹן, דַּי לְכוֹן
‏לְמֶהֱוֵי דָּא עִם דָּא, דְּהָא מִסְכְּנִין לָא יִתְבַּטְלוּן מִן עָלְמָא, וּצְרִיכִין
‏לְגָמוֹל עִמְּהוֹן טִיבוּ. דָּלֶ"ת אִיהוּ מִסְכְּנָא, גִּימֶ"ל גָּמוֹל לָהּ טִיבוּ,
‏לָא תִתְפָּרְשׁוּן דָּא מִן דָּא וְדַי לְכוֹן לְמֵיזַן דָּא לְדֵין.

36. The letters *Dalet* and *Gimel* entered. THEY both also claimed the same thing. He told them also be satisfied with being with each other, because there will always be poor men on Earth, and they should be given a benefactor. The letter *Dalet* is poor, BECAUSE IT IS CALLED *DALET*, FROM POVERTY (HEB. *DALUT*), and the *Gimel* reciprocates as a benefactor (Heb. *gomelet*) TO *DALET*. THEREFORE do not leave each other, and it should suffice you that you sustain one another!

(0) The letter *Bet*

עָאלַת אָת **ב** אָמְרָה לֵיהּ: רִבּוֹן עָלְמָא, נִיחָא קַמָּךְ לְמִבְרֵי בִּי
עָלְמָא, דְּבִי מְבָרְכָאן לָךְ לְעֵילָא וְתַתָּא. אָמַר לָהּ קֻדְשָׁא בְּרִיךְ הוּא:
הָא וַדַּאי בָּךְ אִבְרֵי עָלְמָא, וְאַתְּ תְּהֵא שֵׁירוּתָא לְמִבְרֵי עָלְמָא.

37. The letter *Bet* entered and said to Him, Master of the Universe, may it
please You to create the world with me, because by me You are blessed in
the Upper and Lower Worlds. THIS IS BECAUSE THE LETTER *BET* IS THE FIRST
LETTER OF THE WORD BLESSING (HEB. *BERACHAH*). The Holy One, blessed be He,
replied, But, of course, I shall certainly create the world with you. And you
shall appear in the beginning of the Creation of the world.

(P) The letter *Alef*

38. קַיְימָא אָת **א** לָא עָאלַת. אָמַר לָהּ קֻדְשָׁא בְּרִיךְ הוּא: אָלֶ"ף,
אָלֶ"ף, לָמָה לֵית אַנְתְּ עָאלַת קַמָּאי כִּשְׁאָר כָּל אַתְוָון. אָמְרָה
קַמֵּיהּ: רִבּוֹן עָלְמָא, בְּגִין דַּחֲמֵינָא כָּל אַתְוָון נָפְקוּ מִן קַמָּךְ בְּלָא
תוֹעַלְתָּא, מָה אֲנָא אַעֲבֵיד תַּמָּן. וְתוּ, דְּהָא יְהֵיבְתָּא לְאָת בֵּי"ת
נְבִזְבְּזָא רַבְרְבָא דָא, וְלָא יָאוֹת לְמַלְכָּא עִלָּאָה, לְאַעֲבָרָא נְבִזְבְּזָא
דִּיהַב לְעַבְדּוֹ וּלְמֵיהַב לְאָחֳרָא. אָמַר לָהּ קֻדְשָׁא בְּרִיךְ הוּא: אָלֶ"ף
אָלֶ"ף, אַף עַל גַּב דְּאָת בֵּי"ת בַּהּ אִבְרֵי עָלְמָא, אַתְּ תְּהֵא רֵישׁ לְכָל
אַתְוָון, לֵית בִּי יִחוּדָא אֶלָּא בָּךְ. בָּךְ יִשְׁרוֹן כָּל חוּשְׁבָּנִין, וְכָל
עוֹבָדֵי דְעָלְמָא, וְכָל יִחוּדָא, לָא הֲוֵי אֶלָּא בְּאָת אָלֶ"ף.

38. The letter *Alef* stood outside and did not enter. The Holy One, blessed be He, said to it, *Alef, Alef*, why do you not enter and stand before Me like the other letters? It replied: Master of the Universe, because I saw that all the letters left You without benefaction. So what shall I do there myself? Not only that, but You have already presented the letter *Bet* with this greatest gift of all. And it would not be proper for the Supernal King to take back the gift, which He presented to His servant, and give it to another! The Holy One, blessed be He, said, *Alef, Alef*, even though the world is created with the letter *Bet*, you shall be the first (lit. 'head') of all the letters. My attachments shall be expressed only by you and all calculations and actions of the people shall commence with you. Therefore, all unity shall be expressed by the letter *Alef*!

State of Mind;
Mind and Health;
Big Bang;
Physics;
Quantum;
Cause and Effect;
Wireless Broadcasting;
Dark Lord;
Thought;
Desire to Receive;
Plus and Minus;
Power of *Alef Bet*;
Time, Space, and Motion

CHAPTER 1

THOUGHT
AND MIND

WE ARE LIVING IN A
MATERIAL WORLD
AND I AM A MATERIAL GIRL.

—MADONNA

Our state of mind can create illness or speed us on our way to recovery. Just as no boundary exists between man and the cosmos, so, too, are the mind and body inseparable. When we are ill, it is not merely physical disease that must be treated; a shift in the psyche must also occur. This concept is not a new one. Only recently, however, has it gained the attention and respect of the Western world.

Jerome Frank, professor emeritus of psychiatry at Johns Hopkins School of Medicine, has documented some of the connections between mind and health. Research has shown that the mind can play a major role in curing illness. Psychosomatic medicine, the scientific study of the body's startling ability to heighten its own resistance to infection, has aroused much speculation within the medical community. Indeed, psychic imbalance—the disunity of mindset—may well be at the root of all sickness. What else could possibly account for the fact that the administration of placebos has been followed by the healing of life-threatening diseases?

We intuitively recognize the existence of a connection between mind and body, but it is less easy to perceive the link between mind and the cosmos. The ancient axiom "as Above, so Below" alludes to this cosmic connection. From a kabbalistic perspective, all of Creation is part of the Holy Unity. Thus, it only stands to reason that anything occurring anywhere influences everything else instantaneously. This concept stands in opposition to quantum theory's declaration that we live in a random universe. Quantum theory says there is no explanation for why a certain renegade electron decides to behave erratically, unlike millions of its fellows that have dutifully followed the predictable course. From a kabbalistic point of view, however, the random behavior of an electron is an expression of the free will that was imparted to all things at the time of the great Restriction, or *Tzimtzum*, known to science as the Big Bang.

Physicists today see our universe as constantly expanding. It is thought that once the universe reaches its maximum expansion, it will then contract and collapse. Princeton professor John Wheeler observed that "we can well expect that when the universe collapses, it will start a new cycle, and another universe will leave its track in superspace."[6] One definition of space is the distance between two points, but if the two points collapse into one another, does the space between them cease to exist? If this is the case, then Wheeler's term "superspace" might be better restated as "non-space." Physics, in all its theoretical pondering, seems to have finally reached an understanding of non-dimension, which means that physics—together with all of physical science—has planted the seeds of its own demise. For what is physics, if not the study of the physical world? And what is science if its laws apply only within the narrowest frames of reference? Kabbalah, to the contrary, is in no danger of self-destructing. Kabbalists come to their understanding of the universe with due respect for the material world and with full recognition of its many limitations.

When delivering a lecture in honor of the five hundredth birthday of Copernicus, Wheeler said he had a list of three unsolved mysteries: the mind, the quantum world, and the universe.[7] This declaration demonstrated elevated consciousness. By linking the mind with the cosmos, Wheeler revealed a connection with hidden information. He had tuned in, as it were, to a much-ignored transmission station that broadcasts only Truth, twenty-four hours a day. Albert Einstein was attuned to the same frequency when a sudden flash of elevated consciousness revealed to him that in addition to the three spatial dimensions we know, there is a fourth dimension—time itself.

Mind, the quantum, and the universe—all three, said Wheeler, threaten the clean separation between the observer and observed. It is significant that Wheeler mentioned the mind first among his three mysteries. After all, the universe and its constituent quantum world may exist in physical reality, but like the wheel or any other invention, first they must have existed as thought. Yet the segregation of consciousness and matter was for many years the foundation, and the clarion call, of the modern scientific worldview. Quantum mechanics, however, has demolished the view that the universe sits "out there" while we are objective observers of what transpires within it, safely insulated from the action. We now know that by the very act of observing, the observer affects what he observes, which means that human intention alters the very nature of reality.

This understanding leads us to a reconciliation between quantum theory and Kabbalah—between the science of subatomic physics and kabbalistic metaphysics—because both maintain that ours is a participatory universe. The *Zohar* states that in the days of the Messianic Age, "there will no longer be need for one to request of his neighbor, 'Teach me wisdom.'[8] For it is written: 'One day, they will no longer teach every man his brother, for they shall all know Me, from the youngest to the oldest of them.'"[9] Quantum physics provides only the paving stones along the road to this future. The wisdom of Kabbalah is the path itself.

According to the *Zohar*, the day is nearing when the inner secrets of nature, which have remained so long in hiding, will at last be revealed. This knowledge will enable us to grasp the very essence of all that is within us and around us, and will grant us access to the domain of non-space, providing us with a framework for the comprehension of not only our familiar, observable universe but also of that which lies beyond the range of observation in the realms of the metaphysical. This truth will tease open the supreme enigmas of the origins of man and of the universe—both the how and the why.

Who is this observer-creator of whom we speak? He is none other than man, the initiator of thought. *"Sof Ma'aseh BeMachshavah Tehila,"* declares Rav Solomon Alkabets (1500-1580), medieval author of the hallowed Sabbath song *Lecha Dodi* ("Come, My Beloved"): "All manifestations and actions are merely the results of a priori thought," meaning that the law of cause and effect is valid in the spiritual as well as in the material world.

We have no trouble identifying the things we define as "material." The material existence gave rise to the evolution of all of our senses, and accordingly we can see, smell, taste, touch, and hear material things. But when we get down into the substance of what we perceive as solid, physical reality, we return to the basic building block of nature—the atom with its electrons, protons, and nucleus. And what is an electron? Is it a microscopic bit of solid matter? Not at all! It does not even occupy a specific place in space-time. It might best be described as an oscillating electromagnetic field in space/non-space! Thus, we find that the fundamental component of which the material world is constructed is an illusion, and all that remains as indisputably real is what we, at this moment, are sharing: thought-energy-intelligence, the unique, particular life-form that distinguishes one human being from another and which, in truth, is also the whole of the human being.

Astronomer Sir James Jeans summed up this view when he said, "The universe was looking more like a big thought than a big machine. It may even be that what we think is the real and physical universe is just an interference pattern (an impertinent blip) in the world of thought."[10] Kabbalistically speaking, so it is with man, whose physical body interferes with his thought processes, often completely obscuring them. This interference is the power of the Desire to Receive for Oneself Alone or "Dark Force," which can be discerned as the intelligent life-form of the body, while the energy-intelligence of the soul is the Desire to Receive for the Sake of Imparting.

The universe—including man within it—is an enormous composite of thought. We might be inclined to think that an idea as huge as the universe could only be held in the mind of the Lord, but the Grand Unified Thought of which I am speaking is actually a larger manifestation of the same thought that we are sharing at this moment. The Grand Unified Thought exists in our minds, in our bodies, in everything we experience, taste and touch, see and do. Every phenomenon observed— whether particle, antiparticle, neutrino, or quark—is directed by and acts according to the dictates of a particular intelligence of thought, a thought that has been "quantified" by science as a discrete packet of measurable energy, but which, in fact, is a part of the all-pervading whole.

One Single Light binds together everything in Heaven and Earth. Knowing as we do that both positive and negative polarities are required in nature to complete an electrical circuit, it follows that in polarity with the Light, there must also be a balancing counterforce that causes separation and fragmentation, whether we are speaking of the Big Bang or of human relationships. The Kabbalah teaches that there are, in fact, four fundamental forces at work in the universe. In this book, we will be discussing the nature of these forces and their twenty-two subdivisions.

Because we are conditioned to think of the universe in spatial terms, we tend to conceive of forces or energy fields as operating in space. To approach the true meaning of what the kabbalists reveal, however, we must understand that energy does indeed work in space but is not dependent upon space. We asked earlier whether space exists in a form other than the distance between two points. From the kabbalistic point of view, the answer is yes. Non-distant or empty space exists in thought (if we can conceptualize it), but to make use of the concept, we must define it more completely. Because non-distant or empty space does not exist as the distance between two points, we must now describe it as non-space to differentiate it from a quantity of space that does lie between two points. Kabbalah tells us that the energies at work in nature are independent of time, space, and motion, and can best be thought of as ongoing permutations or states of being. At this stage of our investigation, then, let it suffice to say that all energy is created in thought—as states of mind.

We again stress that everything in existence is thought—from a dinner table to an electromagnetic energy field operating in space or non-space, in time or non-time. The mind of an individual, like that of the Lord in whose image he was created, is not only a place where information is stored, but a place where knowledge itself is created.

Wireless broadcasting networks were not invented in the twentieth century. Consciousness acts upon energy-intelligent thought and transfigures it into energy-intelligent matter. Each mind, through its own unique perceptions, programs a new concept into the universal grid, a concept that is instantly flashed to the minds of all our fellow inhabitants of the universe. The instrument by which consciousness performs this miracle is the Hebrew *Alef Bet*.

The Creator, solely of the Energy-intelligence of Sharing, desired to create a Vessel with which to share His beneficence. At first, He considered creating only one other being who would be equal to Him. But the Lord deemed this idea unsuitable to His intention, which was to create a receiver of His munificence. A new entity with power equal to His own would need nothing from Him. Thus the Lord decided to create, at one time, all the souls that would ever exist in the universe, making each a sturdy Vessel capable of receiving and containing His Light.

So within the full spectrum of the Creator's Thoughts, the universe began to take shape. As any master craftsman does when commencing his work, the Creator very carefully considered the tools He would need to use to fulfill His intention. He began to design these tools out of His imagination, whereupon He experienced a marvelous holographic phenomenon: As He wrenched the images of His tools out of the concealment of His Mind, they began to spring into existence! The Creator saw His work, and He was pleased. The designs fairly leaped out of His consciousness.

Twenty-two sublime tonalities emanated, and it pleased the Lord to use them to fashion His universe. These were the twenty-two emanations, from *Alef* through *Tav*, which have come down to us today as the Divine characters of the Hebrew *Alef Bet*. With these twenty-two characters, the Creator planned His Creation, even as He beheld it in Thought-intelligence in an Endless World, devoid of time, space, and motion, where past, present, and future were undifferentiated aspects of the vast panorama of Now. To each of these emanations He assigned a name and a face, and two thousand years before the Creation described in Genesis, He frolicked with these characters and delighted in them.[11]

Just as the material expression of an idea is already contained within the mind, and the oak tree lies within the acorn, so, too, was the universe included within the Supreme Energy-intelligence that would later manifest as the twenty-two letters of the *Alef Bet*. Fashioned by the very Hand of the Lord, these presences are immutable and can never be destroyed. Their eternal nature was revealed when Moses shattered the physical Tablets containing the Ten Utterances on Mount Sinai.[12] The *Zohar* says: "When the Tablets were shattered, the letters flew upward."[13]

The Lord created all of the souls through these twenty-two emanations, each of whom was a trading partner with Him, as stones hewed from a mountain are of the mountain but are not the mountain itself. Thus, He could derive pleasure from imparting, just as they would find pleasure by receiving.

It was not enough for the Lord that His Vessels should gain pleasure from His giving solely by acceding to His will. He wanted them to freely choose to receive His beneficence, so He fashioned free will—the ability of His Vessels to choose between good and evil, darkness and light. For only when the Dark force was an option could mankind opt for the Light. For this reason, we should never forget the limitations of our independence. Though we are hewn from our Maker and are therefore separate from Him in one sense, we are yet *of* our Maker, which means we are of the interconnected universe; consequently, we are each bound by the effects of our thoughts and actions. This is the price of our freedom.

Creation;
Luminous Emanations;
Material World;
Letter-energies;
Tabernacle;
Adam and Eve;
Shattering of the Tablets;
Lost Ark;
Cain and Abel;
Metaphysical Connection;
Planetary Connection;
Past, Present, and Future;
Coded Messages;
Einstein

CHAPTER 2

ORIGIN AND HISTORY OF THE *ALEF BET*

IN THE BEGINNING
THE LORD CREATED
THE HEAVENS
AND THE EARTH.

—GENESIS 1:1

The first line of Genesis is interesting, though its translation is not entirely accurate. Something indeed was created in the beginning, but it was not Heaven and Earth. We read in the *Zohar* that two thousand years prior to the Creation spoken of Genesis, the all-embracing Creator reflected upon and brought into being the twenty-two energy-intelligences that constitute the master communications system through which all subsystems of energy evolve, and which today are expressed as the letters of the Hebrew *Alef Bet*.

The all-embracing Energy-force of Creation, known in biblical terms as the Creator, is in the Kabbalah lexicon described in terms of the Ten *Sefirot* or Ten Luminous Emanations.

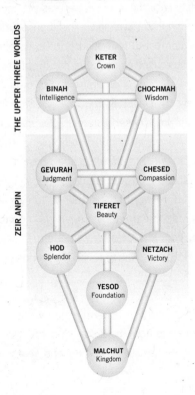

The Upper Three *Sefirot—Keter* (Crown), *Chochmah* (Wisdom), and *Binah* (Intelligence)—are the virtually unknowable Lightforce of the Creator. Of these, only *Binah* is even remotely within the grasp of human understanding. Thus, when the kabbalists speak of the Lightforce of Creation, they are alluding to *Binah*. Keep in mind that *Binah* represents an emanation within the framework of our understanding as it pertains to this physical universe and should not be confused with the higher states of energy known as *Sitrei Torah* (the Secrets of the Bible), which deal with levels of consciousness that are beyond human comprehension.

The *Zohar* declares that when the Creator willed the Creation of the World, the letters of the *Alef Bet* were as yet concealed two thousand years prior to Creation. This concealment of the letters might be compared to a seed prior to its planting. The kabbalistic language used to describe the twenty-two letters is the *Zeir Anpin* microcosm: a grouping of six Lower *Sefirot* plus *Malchut*, the seventh Lower *Sefira*. The six *Sefirot* contained within *Zeir Anpin* are *Chesed* (Mercy), *Gevurah* (Judgment), *Tiferet* (Beauty), *Netzach* (Victory), *Hod* (Splendor), and *Yesod* (Foundation). *Malchut* (Kingdom), or the physical universe, is where any and all of the cosmic energies of *Zeir Anpin* become manifest. The Hebrew word for "world" or "universe" is *olam*; however, the root meaning of *olam* is "concealment." Consequently, when the Creator willed the Creation of the material world, His desire was to *reveal* the worlds of *Zeir Anpin* and *Malchut*.

When we consider all these facts together, we can derive two significant conclusions from the *Zohar*: first, that the twenty-two letters of the *Alef Bet* existed two thousand years prior to the

Creation of the physical universe, and second, that through the emanation of the twenty-two previously concealed letter-energies, the universe as we know it was made manifest.

Further indication of the power of the letters and their combination into words is demonstrated by King David the Psalmist's declaration that "by the word of the Lord were the Heavens made."[14]

The Israelites' construction of the Tabernacle in the wilderness was, and will remain, an incredible feat. How, lacking the proper tools or materials, did they do it? Even with our advanced technology, we can still today only marvel at their accomplishment. The *Talmud* tells us that Bezalel,[15] the ingenious architect, successfully completed the Tabernacle because he "knew how to combine the letters by which the Heavens and Earth were created."[16] The Bible attests to his divinely inspired workmanship by stating: "And He has filled him with the spirit of the Lord, in wisdom, in understanding, and in knowledge, and in all manner of workmanship."[17]

The *Zohar* explains that the addition of a letter from the Lord's Name—the Tetragrammaton: *Yud*, *Hei*, *Vav*, and *Hei* י.ה.ו.ה —to a person's name indicates cosmic intervention. Sometimes, however, a letter condemns a man. We read in the Bible: "And Adam knew his wife, Eve, and she conceived and bore Cain."[18] The energetic impurities of the serpent that seduced Eve are indicated by the first letter of Cain's name, *Kuf* ק. This negative energy, which became manifest within both Adam and Eve, was drawn to Cain by virtue of the letter *Kuf*. Therefore, the internal negative energy of the serpent and the letter *Kuf* are linked each with the other. Consequently, it is written that Adam "knew" Eve, and not that he "begot" Eve. Although Cain's younger brother, Abel, was conceived of the Right, male side, Satan still weakened Adam's power, inasmuch as *Kuf*, the first letter of Cain, became dominant and manifested first. Cain and Abel were two halves of one soul. Consequently, the cosmic influence of Cain's letter *Kuf* passed on to Abel.

As soon as the impurity was eliminated, the letters *Shin* שׁ and *Tav* תּ began to come into manifestation. *Shin* and *Tav* represent the union of male and female. *Shin*, with its three-columned form, is *Zeir Anpin*, or the male aspect, and *Tav* is *Malchut*, or the female aspect. The verse then says: "And he begot a son in his own likeness, after his image and he called his name Seth (Heb. Shet)."[19] שׁ תּ *Shin* and *Tav*, both present within the name Shet, indicate the reunification and manifestation of the all-embracing unified whole through the power of the *Alef* Beth.

Due to their metaphysical nature, the letters of the *Alef Bet* can never be destroyed or mutilated, which is why the letters flew upward when Moses destroyed the stone Tablets bearing the Ten Utterances. The stone Tablets were physical in nature and thus susceptible to destruction, while the energy-intelligences of the letters were metaphysical and indestructible. The Tablets became a broken heap, but the letter-energies removed themselves and remained intact as before. It is written that the Tablets, though fashioned from the hardest sapphire, could be rolled up like a scroll so long as they were connected to the letters—a further attestation to both the interrelatedness of energy and matter and the power of the *Alef Bet*.[20]

Thus, these two Tablets have become a symbol of power, and the mystique surrounding them remains a source of intense curiosity to this day. Indeed, the search for these two Tablets in modern times has provided the background for many stories concerning the lost Ark.

We can find another example of the transcendent power of the *Alef Bet* in the testimonial of the ancient sage Rav Hananiah ben Teradyon. Wrapped in the Scroll of the Bible by the Romans, he cried out, "The parchment is burning, but the letters are soaring on high."[21]

The letters of the *Alef Bet*, when combined and connected to a person's name, render that person capable of achieving altered states of consciousness. This form of intervention via a letter can mean cosmic guidance and protection. After slaying Abel, Cain was toppled from his superior level of consciousness, which caused him to exclaim, "Behold, You have driven me out this day from the face of the land!" ("Face" here denotes a degree of consciousness.) "And from Your Face shall I hide; and I shall be a fugitive and a wanderer in the Earth; and it will come to pass that whoever finds me will slay me." So "the Lord set a sign for Cain, lest anyone finding him should smite him."[22] The protective sign the Lord placed on Cain's forehead, for all to see, was the letter *Vav* ו from His Holy Name—the Tetragrammaton.

When Abram the Patriarch, the world's first and foremost astrologer, saw in his astrological chart that he would never beget a son with his wife, Sarai, the Lord changed his name to Abra_h_am by the adding the letter *Hei* to Abram. Likewise, the Lord removed the letter *Yud* from Sarai's name and placed

Hei—the same letter He used to create Abraham's name—in its stead, making her name Sarah.[23] Here again, the enormous power of the *Alef Bet* was put to a test and engendered a completely altered state of consciousness. Abram and Sarai's natal charts indicated a barren life together without the blessing of an heir, but once they had been raised to a higher level of cosmic consciousness through the intervention of the *Alef Bet*, their natal charts no longer influenced their lives. They were, in effect, transformed into different human beings with a different future.

"And He [the Lord] brought him [Abraham] outside and said, 'Look now toward Heaven and count the stars, if you are able to count them,' and He said: 'So shall be your seed.'"[24] It was a strange dialogue in an abstruse setting. What does observing and counting the stars have to do with the Lord's assurance that Abraham would beget a son? The Lord was assuring Abraham that he would beget a son because of his "going outside," or ascending to a higher level of consciousness.

The kabbalistic interpretation of the now-familiar quotation, "There is no cosmic influence to Israel, for while the stars compel, they do not impel,"[25] emerges from the same biblical reference. The Lord informed Abraham that because he could "count the stars" (that is, have knowledge of the stars) he could transcend the influence of the cosmos and be less subject to planetary influences. The stars compel and influence our actions, but we can still transcend their influence.

Two methods of metaphysical connection exist in our universe. The first is the acquisition of an altered state of consciousness compatible with the larger, all-embracing cosmic reality. In this state, we acquire important knowledge and a necessary awareness of the unseen forces that make up the reality surrounding us, as well as a new comprehension concerning the interrelationships among people, things, and the natural universe. In an altered state, we enter an internal flow pattern that is at-one with the cosmos, whereupon our consciousness becomes connected to the information centers of the cosmos: past, present, and future. "These are the true philosophers, the astrologers of Israel," says the *Zohar*, "who know that which has existed and also know what lies in the future. By virtue of knowing the letters [internal manifestations] of the sun, the moon, and the eclipses of each, the planets, and the signs of the zodiac, they know all that exists in our universe."[26]

The second method of cosmic connection is one where an individual does not have to acquire an altered state of consciousness because he or she is born with an inherent capacity for transcendence. I have coined the phrase "innate consciousness" to explain how some individuals are simply born with an innate mechanism for achieving cosmic awareness.

The seven biblical chariots—Abraham, Isaac, Jacob, Moses, Aaron, Joseph, and King David—personify this second concept. The biblical account of the Seven Days of Creation describes the Lower Seven *Sefirot*, each of which corresponds to one of the seven patriarchs, as follows:

Day One	Abraham	*Chesed*
Day Two	Isaac	*Gevurah*
Day Three	Jacob	*Tiferet*
Day Four	Moses	*Netzach*
Day Five	Aaron	*Hod*

| Day Six | Joseph | *Yesod* |
| Day Seven | King David | *Malchut* |

The word "day" is the cosmic code for *Sefira*, or a life-form-intelligence, as deciphered by the *Zohar*. As has been repeatedly stated, the Bible is a message written in a cosmic code, and the information on how to decipher that code is provided by the wisdom of Kabbalah, and in particular, the *Zohar*.

> *The Heavens declare the glory of the Lord. And the firmament shows His handiwork. Day unto day utters speech, and night unto night reveals knowledge. There is no speech, there are no words, nor is their voice heard. Their line is gone out through all the Earth, and their words to the end of world, in them He has set a tent for the sun.*[27]

The idea that the universe provides us with a coded message has long been known to kabbalists. A major challenge for the world today is to decipher the content of this cosmic code known as the Bible. The kabbalists revealed forces in the universe that could annihilate mankind: "And Moses saw an Egyptian smiting a Hebrew . . . And he looked here and there."[28] Moses knew the secret of the Hebrew letters. "Here and there" refers to Moses looking into the fifty letters by which the Israelites proclaimed the all-embracing cosmic unity by reciting the *"Shema Yisrael"* (Hear Israel) twice a day, and perceiving through elevated consciousness that no good son would ever be born from the Egyptian who was smiting the Hebrew. So, by merely staring at him, Moses brought about the death of the Egyptian through the awesome power of the cosmic force.[29]

From time to time, a true genius with a perfected innate consciousness emerges. Einstein, like all true geniuses, was born with a capacity for cosmic connection. But did he reveal something that had not existed previously? Did he invent anything? Did Einstein's fellow scientists radically alter the state of existence by exploring the structure of the universe? No. Even a genius is not an instigator of new concepts and inventions, as is commonly believed; rather, such a person is a channel for the Cosmic Unity.

At any moment, when cosmic intelligence is ready to reveal itself, someone will be chosen for the job. A particular intelligence revealing an already-existing aspect of our universe will become expressed. The questions of who and why, and why now, are inevitably linked with the concept of reincarnation. Who? Someone who will be able to communicate this new intelligence so that it will be accepted. Why? To provide another link toward ultimate enlightenment. Why now? Based on humankind's activity at the precise time, our cosmic computer system makes contact with the new components and the information is released. In truth, this intellectual energy is born, not from the individual, but from the enormous input of the collective activity of the human race.

Each one of the twenty-two letters is a seed, a starting point on the path toward the Divine Spiritual Consciousness residing in the primacy of four universes/worlds known mnemonically by the acronym *A'BYA*: *Atzilut* (Emanation), *Briah* (Creation), *Yetzirah* (Formation), and *Asiyah* (Action). From these four primal universes emanate the infinite universes that pervade existence.

Words	Sefira	Tetragrammaton	
Adam Kadmon	Keter		
Atzilut	Chochmah	י	
Briah	Binah	ה	Chesed
			Gevurah
Yetzirah	Zeir Anpin	ו	Tiferet
			Netzach
			Hod
Asiyah	Malchut	ה	

The *Zohar*[30] explains that the Lord created two basic realms, each consisting of the four worlds just mentioned. An equal amount of cosmic power of good and evil was granted to these two realms, thus creating the two fundamental systems of good and evil that could now exercise cosmic influence over mankind. Each of the letters of the Hebrew *Alef Bet* is cosmically structured to maintain a balance between good and evil by virtue of their innate spiritual consciousness, considered to be the worthy channel for the Lord's Creation. The battle between good and evil had begun. This cosmic war would continue until the *Kedusha* (Holiness, Purity) triumphed over the *Tumah* (Impurity). Because of this cosmic balance of power between good and evil in our world—the World of Action (*Asiyah*), it is sometimes difficult to distinguish between a truly spiritual individual and one who creates an outward impression of spirituality.

This observation is corroborated by the prophet Malachi:

"And prove me now, said the Lord of hosts, if I will not open for you the windows of Heaven and pour you out a blessing, that there shall not be room enough to receive it. Then you shall return, and discern between the righteous and the wicked, between him that serves God and him that serves Him not."[31]

How can one determine the spiritual worth of an individual? The answer is quite simple. Observe the person's inner peace. Does this person share? Is he or she humble and devoid of hatred? Does he or she take steps to remedy the ills of the world? Does this person display a concerted effort to live by the dictum "love your neighbor as yourself?" Does he or she truly tolerate the opinions and ideas of others? Or does the person exhibit a tendency to force his or her beliefs upon others? By asking and answering these questions, one may find a spiritual mentor.

Each letter of the Hebrew *Alef Bet* is cosmically structured to maintain a balance between good and evil, and each, by virtue of its innate spiritual consciousness, believed itself to be a servant of the Lord and considered itself capable of helping cosmic *Kedusha* dominate cosmic *Tumah*, which would subdue the *klipot* (evil) so that the *Gemar HaTikkun* (Final Correction of the world) could come to pass. The Lord informed each of the letters that, placed alongside their individual cosmic intelligence of *Kedusha*, there existed an equal and opposite cosmic intelligence of *Tumah* granted to the *klipot*. Thus, the Lord could determine which of the innate cosmic powers of the letters was capable of subduing and eradicating evil for all time.

Brain;
Shield of David (Star of David);
Jerusalem;
Patriarchs;
King David;
Age of Aquarius;
Bible;
Golden Calf;
Mount Sinai;
Chariot;
Murder;
David and Goliath;
Batsheva

CHAPTER 3

NUMEROLOGY OF THE COSMIC LETTERS

FOR THE SCIENCE OF
MATHEMATICS IS ONE OF
THE GATES WHICH LEAD
TO A KNOWLEDGE OF THE
SUBSTANCE OF THE SOUL.

—DAVID IBN MERWAN AL-MUKAMMAS

I t is very important to know that the twenty-two letters are divided into three levels of emanation: *Binah* (Intelligence), *Tiferet* (Beauty), and *Malchut* (Kingdom). These three levels of consciousness are subdivided in many ways. Those who have viewed the written Scroll of the Bible have noted that its letters are written in three different sizes. Each size of letter represents a particular intelligent life-form. The larger letters of the Scroll represent the level of consciousness known as *Binah*; the intermediate-sized letters, which constitute the majority of the letters found in the Scroll, reflect the cosmic level of *Tiferet*; and the smaller letters signify the cosmic level of *Malchut*.

Inasmuch as the majority of the letters in the Scroll belong to the cosmic level of *Tiferet*, the *Zohar* refers to the Scroll as *Torah She'Biktav*, the Written Bible. According to the *Zohar*, the Scroll provides access and cosmic connection to the level of consciousness known as *Tiferet*, whereas the *Talmud*, the Oral Bible, provides a connection to the cosmic level of *Malchut*.[32]

The Hebrew letters all have a numerical value. Both numbers and letters are instruments of cosmic energy. In the ones category, there are nine letters, beginning with *Alef* and ending with *Tet*, and these comprise the nine cosmic *Sefirot* of cosmic *Binah*. In the tens group, there also are nine letters, running from *Yud* through *Tzadi*, and these represent the nine *Sefirot* of cosmic *Tiferet*. Within the hundreds category, which relates to cosmic *Malchut*, there are only four letters: *Kuf*, *Resh*, *Shin*, and *Tav*.

Hebrew Letter		Name	Numeric Value	Sefirah
א		Alef	1	Binah
ב		Bet	2	Binah
ג		Gimel	3	Binah
ד		Dalet	4	Binah
ה		Hei	5	Binah
ו		Vav	6	Binah
ז		Zayin	7	Binah
ח		Chet	8	Binah
ט		Tet	9	Binah
י		Yud	10	Tiferet
כ	ך	Kaf	20	Tiferet
ל		Lamed	30	Tiferet
מ	ם	Mem	40	Tiferet
נ	ן	Nun	50	Tiferet
ס		Samech	60	Tiferet
ע		Ayin	70	Tiferet
פ	ף	Pei	80	Tiferet
צ	ץ	Tzadi	90	Tiferet
ק		Kuf	100	Malchut
ר		Resh	200	Malchut
ש		Shin	300	Malchut
ת		Tav	400	Malchut

Why does cosmic *Malchut* contain only four letter-energies? Does the lack of letters indicate a lower cosmic level than that of cosmic *Binah* and cosmic *Tiferet*? The answer is yes.

The nature of the *Sefirot* is illustrated by the symbol of the loosely known term Star of David, the hexagram or six-pointed star formed by two equilateral triangles that have the same center and are superimposed one upon the other but pointing in opposite directions, one upward and one downward. Most people have assumed that the Star of David's sole purpose is a decorative one. Some interpret the Star of David as the planetary sign of Saturn connected with the holy stone in the pre-Davidic sanctuary in Jerusalem. In truth, the hexagram was engraved on the seal or ring of King Solomon as a sign of his dominion over demons.[33] Most scholars believe that its significance as a symbol is connected with the days when King David wore it on his shield. But why did King David choose to use the hexagram on his shield? David was a prophet, not superstitious.

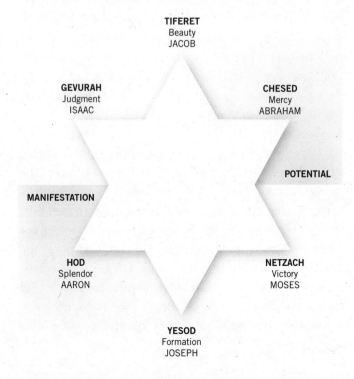

TIFERET
Beauty
JACOB

GEVURAH
Judgment
ISAAC

CHESED
Mercy
ABRAHAM

POTENTIAL

MANIFESTATION

HOD
Splendor
AARON

NETZACH
Victory
MOSES

YESOD
Formation
JOSEPH

> *I see him, but not now;*
> *I behold him, but not near;*
> *There shall step forth a star out of Jacob,*
> *And the Messiah shall rise out of Israel,*
> *And shall smite through the corners of Moab,*
> *And break down all the sons of Seth.*
> —*Numbers 24:17*[34]

The phrase "a star out of Jacob" refers to King David, the first monarch to conquer Moab.[35] The above Bible verse seems to place a great deal of emphasis on the "star." The *Zohar* explains that not only does the six-pointed star have awesome power, but the six points connote the cosmic energy of the six *Sefirot* of *Zeir Anpin*; the *Zohar* then proceeds to describe what role the star will play during the Age of Aquarius. Here we find the key where cosmic energy may become the vital force necessary to cut the world's needless bloodshed, war, and suffering in half. This star also is known as the Shield of David. The *Zohar* says: "King David is the symbol of the Messiah. During the Age of Aquarius, the mysteries of the star shall become revealed to all mankind."[36]

The Upper Triad of the star contains the three *Sefirot* of *Chesed* (Mercy), *Gevurah* (Judgment), and *Tiferet* (Beauty). The chariots, or connecting links, of these *Sefirot* are the patriarchs Abraham (*Chesed*), Isaac (*Gevurah*), and Jacob (*Tiferet*), who made these three energy-intelligences manifest within our universe. Their position within the star indicates the inaccessibility, for the average human being, of the cosmic connection. The Upper Triad is the realm of potential not yet manifested. The power of the Upper Triad is far too great, cosmically speaking, to permit its direct expression. This cosmic energy must be transformed if we are to use it. To achieve the higher consciousness offered by the Upper Triad, we must use metaphysical channels more closely linked with the realm of our universe, the World of Action.

The Lower Triad is structured to provide just such a link. It consists of forces that have moved out of the realm of potential into a manifested state of expression. The Lower Triad contains the three *Sefirot* of *Netzach* (Victory), *Hod* (Splendor), and *Yesod* (Foundation). The chariots, or connecting links, for these *Sefirot* are Moses (*Netzach*), Aaron (*Hod*), and Joseph (*Yesod*). The patriarchs of the Upper Triad—Abraham, Isaac, and Jacob—were not connected during their lifetimes with the people of their generation; their lives were of concealment, privacy, and individuality. But this was not so for the chariots of the Lower Triad—Moses, Aaron, and Joseph. Moses was the leader of his people; Aaron became the High Priest and mediator between the Upper and Lower spiritual realms; and Joseph[37] became second only to the Pharaoh and the provider who delivered the entire known world from disaster and famine. Consequently, it is the Lower Triad through which the world of *Malchut* (our physical world) can connect to the Upper Triad.

The *Zohar* says King David is the chariot for *Malchut*,[38] which became manifest by virtue of David's inherent metaphysical capacity. This is why the Star became known as the Shield of David. It was David who provided humankind with the direct, final link in the chain of cosmic connection. As the *Zohar* states: "In the days of the Messiah, there will no longer be the necessity for one to request of his neighbor, 'Teach me wisdom,' as it is written: 'One day, they will no longer teach every man his neighbor and every man his brother, saying know the Lord. For they shall all know Me, from the youngest to the oldest of them.'"[39] The Age of Enlightenment thus appears to be strongly linked with the Age of Aquarius.

As previously mentioned, the Upper Triad of *Tiferet* is inaccessible to us, for it is beyond the reach of human consciousness. Our world, the World of Action, can only connect with the Lower Triad and with *Malchut*, which embodies only four letter-energies: *Kuf, Resh, Shin,* and *Tav*. Thus, our link to the Upper Levels of consciousness must come about through these four letter-energies.

The Bible immortalizes a galaxy of great men and women who made significant contributions to the Israelites and to the well-being of all people throughout the world. There is, however, one individual who towers above them all: Moses. The Bible attests to his unique power and importance when the Lord declares, "Hear now My words; if there be a prophet among you, I the Lord do make Myself known unto him in a vision, I do speak with him in a dream. Not so with My servant Moses; he is trusted in all My House. With him, I speak mouth to mouth and not dark speeches."[40]

Not only was Moses the first teacher of the nation of Israel, but he also gave the mystical energy of Bible to the world. Despite many setbacks, he never faltered. Always he persevered. It was Moses who transformed a horde of slaves into a nation with the potential to secure peace and harmony in the world. Even when Israel fell under the influence of the golden calf, Moses did not abandon his people. Although this tragic event left a spiritual wound so deep that Moses shattered the Tablets written upon by the very Hand of the Lord, he nevertheless intervened on behalf of his flock in a heart-rending prayer: "Yet now, if You will, forgive their sin; and if not, blot me out, I pray You, out of Your Book which You have written."

Why was Moses chosen to lead the Israelites? Moses was so kind to Jethro's flock that the Lord Himself is said to have remarked, "If Moses tendered such loving care to animals, then how much greater would his love for humans be."[41] Moses was a supreme humanitarian, capable of playing the role of intermediary. The important letter *Kuf*, examined in a later chapter, is a symbol of the link between Moses and the Upper Triad. From his position within the cosmic *Sefira* of *Netzach*, Moses, through the letter *Kuf*, provides a symbolic bridge to the highest states of consciousness.

Aaron, brother of Moses and Miriam, belonged to the tribe of Levi.[42] While the biblical narrative clearly assigns Aaron a role subordinate to that of Moses, Aaron undoubtedly held a high position within the hierarchy of his tribe. The Lord often communicated with Moses and Aaron together.[43] Aaron was able to make atonement for the people through incense-offerings.

Aaron is especially praised for his love of peace. He never disgraced people by telling them they had sinned. He employed every stratagem possible to reconcile disputes, particularly between man and wife.[44] Some believe his love for his fellow man determined his attitude regarding the golden calf. It was during Moses' prolonged absence on Mount Sinai that Aaron yielded to the people's demand and fashioned a golden calf, which the Israelites then deified.[45] He could have put those who worshipped the golden calf to death, as Moses later did, but his love and compassion for his people prevented him from executing this decree.

Aaron considered peaceful persuasion the best means for an appreciation of the Bible. Hillel said, "Be like the disciples of Aaron, loving peace, loving one's fellows, and bringing them closer to the Bible."[46] No direct reason is given for the selection of Aaron as High Priest and founder of a hereditary priesthood, but the *Zohar* connects Aaron's love of peace to his elevation as High Priest.[47]

Originally, Aaron was the chariot or metaphysical link with the *Sefira* of *Hod*. When he was spiritually prepared to ascend to the higher level of consciousness known as *Chesed*, that echelon of priesthood consciousness became accessible to him and he was able to embrace cosmic energy more powerful than was possible for any other Levite.

When Aaron's spiritual activity became isolated from the mainstream of his society because of his spiritual elevation, Korach and his cohorts led a challenge to the monopoly held by Aaron's family on the priesthood. The exclusive right of Aaron's family to the priesthood was upheld in a trial by ordeal that led to the destruction of Korach and most of Tribe of Korach and its rebels.[48]

Joseph the Righteous is a chariot of the *Sefira* of *Yesod* (Foundation) in the Lower Triad of the Star of David.[49] The son of Jacob and Rachel, Joseph was born in Paddan-Aram after Rachel had been barren for seven years.[50] Jacob favored Joseph over his children by his other wives. The kabbalistic interpretation of the close bond between this father and son is decidedly different from what the biblical story might seem to imply. According to the wisdom of Kabbalah, the mystery of their affinity consisted of their positioning as a cosmic link to the power of the mediating principle known in Kabbalah as the Central Column. Jacob is the chariot of *Tiferet*, the Central Column of the Upper Triad of the Star of David, while Joseph is the chariot of *Yesod*, symbolizes the Central Column of the Lower Triad. Thus, Jacob and Joseph are both considered to be directly responsible for connecting the celestial Central Column with our terrestrial universe.[51]

From a literal interpretation of the Bible account of Joseph's youth, we learn that he became completely alienated from his brothers because he reported their misdeeds to their father.[52] However, according to the *Zohar*, this estrangement between Joseph and his brothers symbolizes Joseph's elevation to a higher level of consciousness. Joseph rose to the level of *Yesod* of the Lower Triad of the Star of David, while the consciousness of his brothers remained within the influence of the twelve signs of the zodiac, this physical world.[53]

On one occasion, Jacob sent Joseph to visit his brothers and to report back on their welfare. When the brothers saw Joseph from a distance, their hatred and hostility toward him was said to have been so strong that they wanted to murder him.[54] Fratricide is an unsettling proposition, but if we assume the entire Bible to be a cosmic code, as the kabbalists do, then this biblical account of cold-blooded murder takes on a deeper significance. Kabbalah teaches that the sons of Jacob symbolize the twelve signs of the zodiac; therefore, the story of Joseph and his brothers becomes an account of cosmic influences and their sway over the universe.

Joseph was sold into slavery to man named Potiphar, courtier and chief steward of the Pharaoh. Joseph earned the confidence of his master, but his good fortune did not last long. Potiphar's wife attempted to seduce Joseph, who resisted her. Angered by his rejection of her, she slandered Joseph and had him thrown into prison.[55] Here again, the kabbalistic interpretation of this sad episode in Joseph's life differs markedly from a conventional reading of the biblical story. From the kabbalistic perspective, Joseph's sorry state of affairs was a blessing in disguise because it provided him with a new level of awareness, which was necessary to his becoming the chariot of *Yesod*.

Joseph's elevation to cosmic consciousness engendered a new series of strange events. Joseph was brought to the palace to interpret the Pharaoh's dreams, whereupon he predicted seven years of great abundance in Egypt, followed by seven years of famine. Joseph offered advice on how the Pharaoh might deal with the situation, and it was no accident that the Pharaoh adopted his suggestions and appointed Joseph, at age thirty, to be in charge of their implementation.[56]

Joseph's policies resulted in sustenance, rather than famine, for the entire known world, a fact that corresponds precisely to Joseph's position within the Star of David. As the chariot of *Yesod*, Joseph provides the direct cosmic link between the Upper Worlds of *Zeir Anpin* and the *Sefira* of *Malchut* (our physical world). When we decode the biblical narrative from a kabbalistic perspective, we see that it describes Joseph's connection to this world, while also pinpointing his position within the Lower Triad of the Star of David.

David was the youngest son of Jesse of the Ephrathite family who lived in Beth-Lechem in Judah.[57] According to the Book of Ruth, David was descended from Ruth, the Moabite wife of Boaz.[58] David reigned as king of Israel for forty years. His reign consisted of two periods. For seven years, King David reigned over Judah from the city of Hebron. During the second period, he ruled over all of Israel from his capital city of Jerusalem.

David, however, is more popularly known for his heroism against Goliath, the Philistine. This feat, together with David's other successes in the long war with the Philistines, caused him to rise to a high rank in the army of Saul, who was then king of Israel.[59] King Saul subsequently chose David to marry his daughter. But eventually Saul became resentful of David's rising popularity, his jealousy grew into a suspicion that David was plotting to seize the throne. When David realized that Saul was determined to kill him, he fled into exile, where he lived in constant fear of Saul and his army. During this period, David gained the admiration and deep friendship of Saul's oldest son, Jonathan.[60]

By all the laws of human nature, Jonathan should have envied and hated David, yet he loved David more than he loved himself, despite the fact that Jonathan's loyalties were divided between his father and his friend. The biblical account of Jonathan's friendship with David is expressed in the verse: "The soul of Jonathan was knit with the soul of David, and Jonathan loved him as his own soul."[61] David reciprocated Jonathan's love in full measure, for he was unencumbered by the strains and conflicts that tested Jonathan's love. David's love for Jonathan continued uninterrupted despite his precarious situation. Most people in David's place—exiled and fearing for their lives—would have made self-preservation their uppermost priority. But not David, who refused to abandon those whom he loved. No other person in the Bible is described as more noble in his humility and self-denial than David.

King Saul's death in the war against the Philistines ushered in an important stage in David's life. After receiving the news of Saul's death, David understood that the way was clear for the fulfillment of his destiny: He was now the only person in Israel endowed with the military genius and public esteem necessary for the slow and painful task of rebuilding and making manifest the Kingdom of the Lord, *Malchut*.

David took a crucial step after he ascertained the Lord's will by means of the *Ephod*, an ornamented vestment worn by the High Priest over a blue robe. This breastplate was the principal tool for communication with the level of consciousness of the Lord. The coded term for consulting the *Ephod* is "to come before the Lord."[62] David's consultation of the *Ephod* is described thus: "And it came to pass that David inquired of the Lord, saying, 'Shall I go up into any of the cities of Judah?' And the Lord said unto him, 'Go up.' And David said, 'Where shall I go up?' And He said, 'Unto Hebron.'"[63]

David's endeavors to unify the different tribes into the kingdom of Israel provoked tension and animosity. In the ensuing tumult, his own son Absalom rebelled against him. The Benjamites expressed their enmity toward David by saying, "We have no share in David, and we have no portion in the son of Jesse."[64] King David nevertheless succeeded in preventing the fragmentation of Israel, transforming the Israelites into one centrally governed nation. He led a rich, fascinating life, but the unification of his kingdom towers over his other accomplishments. The *Midrash* declares that the Lord "looks forward to David being king until the Age of Aquarius."[65] The *Talmud* takes a more forceful position concerning the Kingdom of David, saying: "Whosoever contends against the sovereignty of the House of David deserves to be bitten by a snake."[66]

The story of David and Batsheva is a love story. It reveals the mysteries of soul mates and illustrates the uncanny effect that love can have on the lives of men and women.[67] If viewed from the kabbalistic perspective, this story becomes a metaphor that sheds new light on the meaning of love.[68] Hollywood and history say that King David committed adultery with Batsheva and arranged for the death of her husband, Uriah. But according to the *Talmud*, King David was guilty of neither crime. While the text seems to imply criticism of King David's behavior with Batsheva, the general tendency is to exonerate him from all blame because he and Batsheva, as soul mates, shared a love made in Heaven. "Whoever says King David sinned is mistaken," says the *Talmud*.[69] King David contemplated the act, but did not go through with it. Kabbalah thrives on paradox, and the story of David and Batsheva is paradox personified.

We perceive something of a contradiction between King David's preoccupation with war and his involvement with love and poetry. The rabbis speak in superlatives of King David's poetic genius: "All the praises stated in the Book of Psalms were uttered by David,"[70] one says. "While still dwelling in his mother's womb, he recited a poem," says another.[71] "Midnight never found me asleep," says King David of himself. Each night until midnight, he studied the Bible; thereafter, he recited songs and praises.[72]

King David's descent from Ruth the Moabite is echoed in the esoteric teachings. King David, the first Messiah, descended from an admixture with the *klipot*, as will the last Messiah. King David acquired the chariot of *Malchut* (Kingdom) to act with mercy and charity, and yet also to kill according to the law. Of necessity, he enlisted the power of evil to overcome the evil rooted in man. King David seemed to manifest just about everything our society has to offer. "As a counterpart to the biblical David, the Lord has another David," states the *Zohar*.[73] Together with Abraham, Isaac, Jacob, Moses, Aaron, and Joseph, David is the *merkavah*, or chariot, of this World of Action. David combined the two Worlds, Upper and Lower. He was the symbol of Kingdom, the totality of our universe.

The life of David portrays the very essence of mankind's existence. The ongoing battle in the Kingdom symbolizes our inner conflict. Thus, it is fitting that David be included among the Lower Triad of patriarchs because he embraced all of humanity.

Desire to Receive;
Restriction;
Plus and Minus;
Paradox;
Bread of Shame;
Resistance;
Cosmic Communication;
Astronomy;
Astrophysics;
Newtonian Physics;
Why and How;
DNA;
Genetic Codes;
Seven Planets;
Human Body

CHAPTER 4

THE HEAVENLY CONSTELLATIONS

NO POINT IS MORE
CENTRAL THAN THIS,
THAT EMPTY SPACE
IS NOT EMPTY.
IT IS THE SEAT OF THE
MOST VIOLENT PHYSICS.

—JOHN A. WHEELER

When God willed the Creation of the world, all the letters of the *Alef Bet* appeared before Him in reverse order. Hence, the last letter, *Tav*, was the first to appear. Why did the letters appear in reverse order when the time came for the Will of the Creator to be physically expressed? The reverse order signified that the Desire, or Will, to Receive within the Endless World underwent Restriction, or *Tzimtzum*, which is explained as follows:

> *Since the Desire to Receive, which had been established in the Ein Sof, was receiving the infinite beneficence of the Creator continuously, there arose a feeling called Bread of Shame. The Vessel was receiving but could do nothing in return, inasmuch as the Creator, being whole and lacking nothing, has no Desire to Receive. Thus, the Vessel felt Bread of Shame because it was unable to earn what it was receiving. The metaphysical energy generated by this situation manifests itself as a feeling of guilt, which leads, by the principle of cause and effect, to a voluntary shutting-off of the Light, so that the Vessel can redress the lack of balance caused by receiving that which is not earned. The essential laws regarding the flow of energy in our universe were therefore established in the process of Creation.*[74]

Thus, the paradox of opposites was born. A circuit for cosmic energy was created through an interplay of opposing forces. At first glance, this might seem to be a contradiction in terms; the idea of an intrinsic unity of opposites seems foreign to our way of thinking. However, from a kabbalistic point of view, opposites are simply different aspects of the same all-embracing Unity.[75] Inhaling necessitates exhaling; receiving requires rejection/restriction. For a light bulb to light up, the filament must return the current that is invited by the negative pole. To receive directly without first returning or rejecting the energy sought is to invite a short circuit.

The rejection of energy creates the intelligence referred to in Kabbalah as *Or Chozer* (Returning Light). This newborn intelligence takes the place of the negative pole and acts as a Vessel. Paradoxically, an intelligence created for the purpose of rejecting the Light now assumes the role of receiving the Light. This essential law, which functions both in the physical and metaphysical realms of existence, emerged from the process of Creation. Once established in the metaphysical web of connection, this new intelligent life-form (Returning Light) allows for the removal of Bread of Shame and permits the flow of cosmic energy to be reestablished.

This fundamental principle of resistance is the basis for the twenty-two letters appearing before the Creator in reverse order—from *Tav* to *Alef*. All cosmic phenomena must behave in accord with the paradoxical law of Returning Light. The letter-energies, which are the ultimate in cosmic communication, could not disregard the principle of Returning Light, and so the procedure for the establishment of the world conformed to the laws and principles illustrated by the concept of Bread of Shame. Nothing can violate the principle of universal restriction or resistance. The path of least resistance would merely result in a short circuit. Consequently, when the Creator sought to create the world with the twenty-two intelligent energies, their own intelligence dictated that their appearance before the Creator should be in reverse order.

The sky we see on a clear night has changed very little since the days of Aristotle Ptolemy, Shabbetai Donolo, and the Rabad (Rabbi Abraham ben David). The fact is that it isn't the stars that have changed, but rather our thirst for a rational understanding of them. Yet the more we learn about how the heavenly bodies came into existence, what they are made of, and how they live and die, the more we are filled with wonder and surprise.

The saga of modern astronomy has been one of extracting more and more data out of each light beam and cosmic ray. Diligent research over the ages has brought to light mysteries that were once hidden. A single lifetime, even if it were entirely dedicated to the comprehension of the heavens, would not be enough for the exploration of so vast a subject. Still, by performing some of the most remarkable detective work in modern times, astrophysicists have been able to piece together a dossier on the history of our universe so that we now have some idea of how universes form, live their lives, and ultimately die.

Newtonian physics separated and thus alienated man from the universe. Astronomy revealed a universe so vast and ancient that humanity seemed to be of little or no consequence. Were it not for the ancient science of astrology, the thin connective strand between the vast universe and tiny humanity might have been severed completely. Today, however, quantum science offers new perspectives for understanding the cosmos, and astrophysicists are rediscovering mankind's intimate link with the cosmos. Again we see that the most fundamental natural laws, along with seemingly trivial acts of man, can be traced back to the heavens and their origin.

Humankind has been provided with universal truths that guide us through the illusion of physicality to the ultimate all-embracing reality. This primal life-flow is directed and filtered through many diverse levels of consciousness, establishing infinite layers of energy-intelligences. Yet all phenomena are but one single Lightforce expressing itself in an endless multitude of forms and patterns. All energy and intelligent life-forms come from this one basic Lightforce.

The Lightforce reveals itself though a continuum of events through which man may experience the wellspring of Creation. In all of Creation, there always appears a repetition of the same design:

> *There is not a member of the human body that does not have a counterpart in the world as a whole. For as man's body consists of members and parts of various ranks, all acting and reacting upon each other so as to form one organism, so does the world at large consist of a hierarchy of created things. When they properly act and react upon each other, together they form one organic body. The Bible contains all of the deepest and most recondite mysteries. All essences, both of the Higher and the Lower grades of this world and of the World to Come, are to be found there.*[76]

What seems to emerge from the *Zohar* is a startling revelation regarding the existence of a single pattern in all structures, from the minutest atom to the vastest reaches of the cosmos. The immensity of our universe is revealed in the microcosm. One set of universal laws pervades and unifies all of the intelligent life-forces in the cosmos. These intelligent energies physically manifest within four classes of Creation: mineral, vegetable, animal, and human.[77]

Modern physics has left us with a fear of the cosmos blowing apart forever. Some people even think the principles that govern the universe are like cards randomly shuffled and reshuffled. Others believe that our natural laws represent only a fracton of an infinite range of possible natural laws. The kabbalistic worldview, then, should be welcomed by the Earth's theory-weary inhabitants. Fear not, for the balance of nature is maintained by immutable universal laws.

The *Zohar* states that there is an infinite hierarchy of universes. Even an elementary particle, once penetrated, will reveal an entire galaxy within itself. This is what the *Zohar* alludes to when it says: "For there is not a member of the human body that does not have its counterpart in the world as a whole." Astrophysicists hope that space telescopes will revolutionize our understanding of the cosmos. But if we really want to comprehend infinity, let us now resolve to forget the telescope, for everything and anything included within the universe is infinite. We are infinite. The infinite is within.

Billions of years ago, a cataclysmic event known as the Big Bang gave rise to the physical universe.[78] Theories abound as to *how* the Big Bang happened, but most scientists have yet to ask the question of *why* it happened. From the kabbalistic perspective, we have not penetrated to the heart of a matter until we know *why* it happened. The failure of the scientific imagination to comprehend the crucial significance of the *why* has resulted in the false perception that the only questions worth asking are the ones that science can answer. Modern science has achieved many dramatic successes, but it would be foolish to suppose that the fundamental questions concerning the purpose of the universe have been answered by scientific discoveries.

Paranormal experimenters claim that the human mind can exert force on distant matter, but these assertions have yet to be scientifically validated. It is the latest scientific theories and discoveries concerning the basic oneness of the universe that are the most significant revelations of modern physics. As science penetrates more deeply into matter, descending further and further into the realm of subatomic particles, we see the same cosmic unity expressed over and over again, thus confirming what metaphysicians have known for thousands of years: The universe is one living, breathing organism.

Abraham the Patriarch's statements in *Sefer Yetzirah* (*Book of Formation*) are a living testimonial that the constituents of matter—the cosmos, man, space-time, and the primary phenomena involving them—have evolved from one basic unifying force. The cosmic bond holding all these phenomena together are the Hebrew letters of the *Alef Bet*.

The letter-energies, then, should not be viewed and understood solely as distinct and isolated entities. Our fragmented view of the universe is merely a manifestation of our own internal fragmentation. In fact, we are all connected with the whole vast space-time continuum and everything in it. The physical world and the internal world really are two sides of the same fabric. The threads of all events and intelligent life-forces are woven into a harmonious web of infinite interactions and connections.

What emerges from the *Book of Formation* is a case for the existence of a superior intelligence. This stands in contradiction to the standard scientific position, which attributes the physical qualities of an organism solely to the structure of its DNA. The first evidence for this latter stance came from an experiment conducted by Oswald Avery in 1946, which showed that only DNA transmits hereditary properties. We can modify other molecules, such as proteins in an organism, but these changes will not be transmitted to subsequent generations. However, when DNA is modified, all successive generations of that organism will be affected. Since that time, the overwhelming weight of scientific evidence indicates that the only way of changing the instructions for building a new organism is to alter the DNA itself, which further implies that those instructions must, of necessity, be coded somehow in the structure of the DNA.

Similarly, kabbalists tell us that the *Alef Bet* contains the code responsible for the creation and development of everything contained within our universe, micro and macro. Both worlds emanated from the same intelligent life-energy, which modern science has overlooked in its preoccupation with reducing systems into smaller and smaller fragments. Abraham, in the *Book of Formation*, declared the universe to be one cosmic whole. Just as the DNA molecule carries the entire genetic code of a being, so, too, does the Hebrew *Alef Bet* contain the entire code of our universe. The *Alef Bet* was the primordial ancestor of DNA, the master molecule of life on Earth. In fact, the discovery of DNA was the inevitable outcome of centuries-old kabbalistic work that is still being carried on in our own lifetimes. Indeed, the unraveling of the molecular structure of DNA in the early 1950s by Francis Crick corresponds to ideas presented in the *Book of Formation* by Abraham the Patriarch. Through the *Alef Bet*, we can make contact with the true components of the cosmic code—the permanent natural laws of our universe. While the *Alef Bet* embodies the coded instructions to replicate itself in subsequent words, the origin of the instructions, or the mechanisms that produce the genetic codes, remains a mystery.

What is the origin of the Earth, the moon, and the sun? What are the origins of the galaxies and the cosmos as a whole? Who or what ordered them into being? Why was the order given?

While the scope of this book does not address itself extensively to these questions, the explanation of the Big Bang found in the *Zohar*—a book written some two thousand years ago—does describe in detail the framework of the original Creation. For most people in the Western world, our initial exposure to these questions comes in the biblical account of Creation in Genesis.[79] However, this account gives only a vague description of what happened, although the *Zohar* tells us that this story—and indeed, the entire Bible—can be read as a cosmic code. For the uninitiated reader, the *Zohar*'s decoding of the Bible is even more abstruse than the Bible itself, but fortunately Rav Ashlag, one of the greatest kabbalists of this century and founder of the Kabbalah Centre in 1922, was able to explain those parts of the *Zohar* that revealed the precise natural order displayed in Creation and in the Heavens. Consequently, from seemingly insignificant biblical tales, we can derive the *Zohar*'s interpretation of the Big Bang.

Let us examine more closely the Bible verse that the *Zohar* considers to be the basis for the Creation of the world. Strangely enough, the passage is not included in the original description of Creation: "And Melchizedek, king of Salem, brought forth bread and wine."[80] The *Zohar* says this is written as: "The letter-energy *Hei* [*Malchut*] was crowned with the letter *Vav* [*Zeir Anpin*], and *Vav* with *Hei*; thus *Hei* ascended and was joined in a perfect cosmic bond. This is alluded to in the words, 'Melchizedek [literally, King of Righteousness], king of Salem [literally, completeness],' that is, the king who rules with complete sovereignty."[81]

The *Zohar* describes the Creation of the world in this way:

> When the Lord decided to create the world, He first produced a flame from a scintillating lamp. He blew spark against spark, causing darkness and fire, and produced from the recesses of the abyss a certain drop which He joined with the flame, and from the two, He created the world. The flame ascended and encircled itself with the Left, and the drop ascended and encircled itself with the Right. They then crossed and changed places, going up and down alternately until they were closely interlocked. There issued from between them a full wind. Then those two sides were made one, and the wind was set between them, and they were entwined with one another. And so there was harmony Above and harmony Below.

On the basis of this and other sections of the *Zohar*, Rav Isaac Luria, the great sixteenth century Kabbalist known as the Ari, and Rav Ashlag confidently answered the question of where the universe comes from: "It came out of a vacuum. The entire physical universe is a re-expression of sheer nothingness."[82]

While all intelligent energies must, of necessity, have been included within the vast scope of the Endless World, they became physically expressed through the letter-energies. Which letters were responsible for making manifest this cosmic energy? The *Zohar* answers: "From the Living Energy-Intelligence (Spirit) emanated Air (*Ruach*); from Air, Water; and from Water, Fire."

Of the twenty-two letters or inherent powers, three represent the first elements. These three fundamental elements were manifested by the letters *Mem*, *Shin* and *Alef*, and form the basis for balance:

The *Mem* מ is mute like Water, and it makes manifest.

The *Shin* ש hisses like the Fire it brings forth.

The *Alef* א is a breath of Air that reconciles the other two.

Heaven was created from Fire, the Earth (consisting of sea and land) from Water. Air, the atmosphere, establishes the balance between them. The three fundamental letter-energies *Shin*, *Mem*, and *Alef* caused heat, cold, and moisture. Heat was created from fire, cold from water, and moisture from air, which equalizes them.[83]

Next came *Bet*, *Gimel*, *Dalet*, *Kaf*, *Pei*, *Resh*, and *Tav* that were designed, established, combined, weighed, and changed by the Lord. With these letters, He formed the seven entities in our solar system that the ancient astronomers described as Saturn, Jupiter, Mars, the Sun, Venus, Mercury, and the Moon. The Lord also used these letters to create the seven days of the week, as well as the "seven gates" (the seven openings of the senses in the human body), which are the two eyes, the two ears, the two nostrils, and the mouth.

Hebrew Letter	Hebrew Letter	Element	Planet	Body Part	Constellation
א		Alef	Air			
ב		Bet		Saturn	Right Eye	
ג		Gimel		Jupiter	Left Eye	
ד		Dalet		Mars	Right Ear	
ה		Hei			Right Hand	Aries/Nissan
ו		Vav			Left Hand	Taurus/Iyar
ז		Zayin			Right Leg	Gemini/Sivan
ח		Chet			Left Leg	Cancer/Tamuz
ט		Tet			Right Kidney	Leo/Menachem Av
י		Yud			Left Kidney	Virgo/ Elul
כ	ך	Kaf		Sun	Left Ear	
ל		Lamed			Liver	Libra/Tishrei
מ	ם	Mem	Water (Earth)			
נ	ן	Nun			Spleen	Scorpio/ Mar Cheshvan
ס		Samech			Gall Bladder	Sagitttarius/ Kislev
ע		Ayin			Stomach	Capricorn/Tevet
פ	ף	Pei		Venus	Right Nostril	
צ	ץ	Tzadi			Intestines	Aquarius/Shevat
ק		Kuf			Stomach	Pisces/Adar
ר		Resh		Mercury	Left Nostril	
ש		Shin	Fire (Heaven)			
ת		Tav		Moon	Mouth	

The Lord infused the letter *Bet* with an intelligent energy-force to predominate in wisdom, and with it, He formed the planet Saturn in the cosmos, the first day of the week, and the right eye in man.

The Lord infused the letter *Gimel* with an intelligent energy-force to predominate in peace, and with it, He formed the planet Jupiter in the cosmos, the second day of the week, and the left eye in man.

The Lord infused the letter *Dalet* with an intelligent energy-force to predominate in war, and with it, He formed the planet Mars in the cosmos, the third day of the week, and the right ear in man.

The Lord infused the letter *Kaf* with an intelligent energy-force to predominate in regality, and with it, He formed the Sun in the cosmos, the fourth day of the week, and the left ear in man.

The Lord infused the letter *Pei* with an intelligent energy-force to predominate in love, and with it, He formed the planet Venus in the cosmos, the fifth day of the week, and the right nostril in man.

The Lord infused the letter *Resh* with an intelligent energy-force to predominate in the arts, and with it, He formed the planet Mercury in the cosmos, the sixth day of the week, and the left nostril in man.

The Lord infused the letter *Tav* with an intelligent energy-force to predominate in the Kingdom, and with it, He formed the Moon in the cosmos, the seventh day of the week, and the mouth of man.[84]

The remaining twelve letters were each provided with an intelligent energy-force to predominate the twelve signs of the zodiac, and with these letters, the Lord also established the twelve months of the year and the twelve leaders in the human body, male and female. The twelve constellations of the zodiac are Aries, Taurus, Gemini, Cancer, Leo, Virgo, Libra, Scorpio, Sagittarius, Capricorn, Aquarius, and Pisces. The twelve months of the year are *Nissan*, *Iyar*, *Sivan*, *Tammuz*, *Menachem Av*, *Elul*, *Tishrei*, *Mar Cheshvan*, *Kislev*, *Tevet*, *Shevat*, and *Adar*. The twelve leaders of the human body are the two hands, the two feet, the two kidneys, the spleen, the liver, the gall bladder, the stomach, the upper intestine, and the lower intestine.

The Lord infused the letter *Hei* with an intelligent energy-force to predominate in Aries in the cosmos, *Nissan* in the year, and the right hand in man.

The Lord infused the letter *Vav* with an intelligent energy-force to predominate in Taurus in the cosmos, *Iyar* in the year, and the left hand in man.

The Lord infused the letter *Zayin* with an intelligent energy-force to predominate in Gemini in the cosmos, *Sivan* in the year, and the right leg in man.

The Lord infused the letter *Chet* with an intelligent energy-force to predominate in Cancer in the cosmos, *Tammuz* in the year, and the left leg in man.

The Lord infused the letter *Tet* with an intelligent energy-force to predominate in Leo in the cosmos, *Menachem Av* in the year, and the right kidney in man.

The Lord infused the letter *Yud* with an intelligent energy-force to predominate in Virgo in the cosmos, *Elul* in the year, and the left kidney in man.

The Lord infused the letter *Lamed* with an intelligent energy-force to predominate in Libra in the cosmos, *Tishrei* in the year, and the liver in man.

The Lord infused the letter *Nun* with an intelligent energy-force to predominate in Scorpio in the cosmos, *Mar Cheshvan* in the year, and the spleen in man.

The Lord infused the letter *Samech* with an intelligent energy-force to predominate in Sagittarius in the cosmos, *Kislev* in the year, and the gall bladder in man.

The Lord infused the letter *Ayin* with an intelligent energy-force to predominate in Capricorn in the cosmos, *Tevet* in the year, and the stomach in man.

The Lord infused the letter *Tzadi* with an intelligent energy-force to predominate in Aquarius in the cosmos, *Shevat* in the year, and the intestines in man.

The Lord infused the letter *Kuf* with an intelligent energy-force to predominate in Pisces in the cosmos, *Adar* in the year, and the stomach in man.[85]

The *Sefer Yetzirah* (*Book of Formation*) represents a theoretical approach to the problems and answers of cosmology and cosmogony. The book is distinguished by its brevity, with even the most comprehensive of the numerous editions not exceeding seventeen hundred words. This work is the earliest existing text written in the Hebrew language. Its chief subjects are the elements of the entire cosmos and its occupants—the master code, or DNA, of all Creation. The twenty-two letters of the Hebrew *Alef Bet* together exhibit the mysterious forces whose convergence produces the combinations of entities and events observable throughout the whole of Creation.

Before the Big Bang;
Vast Emptiness;
Evil;
Process of Correction of the Soul;
Death;
Sin;
Knowledge;
Universal Consciousness;
Dark Side;
Master of Evil;
Seal of Death;
Judgment

CHAPTER 5

THE LETTER
TAV

BY BEING AWARE OF THE
DIFFICULTY OF A PROBLEM,
WE ARE GUIDED TO
THE WAY WHICH LEADS US
TO THE ATTAINMENT OF
THE TRUTH THEREOF.

—GERSONIDES

The cosmos was vast and empty, an infinite silence teeming with Light, thought, and consciousness, but devoid of matter. The Big Bang was still an event that would occur in what man would have called the distant future—had man been present to speculate upon the point. Not even the dust of Adam's molding was present yet, but out of that awesome void, man and the universe—cosmic embryos in the Mind of the Lord—were about to be born. And one—just one—of the twenty-two living letters of His beloved *Alef Bet* would become the channel of Creation.

Before the beginning, there existed only infinity. Everything was simple and smoothly balanced in one likeness, measurable only by the energies that traveled through dark holes in the innermost recesses of Endlessness.[86] There, in the cosmic thoughts of *Binah*, *Tiferet*, and *Malchut*, the letters marked time in the universe. Stars collapsed, supernovae sped across the vast emptiness, and intelligences were formed and then died. Yet the letter-energies knew that the moment of Creation would come. Thus, eager to be chosen as the vehicle for that staggering outflow of energy, each came before the Lord to plead her case (Hebrew letters are feminine), and to state her merits and the qualifications by which she should be chosen as the channel for the birth of Creation.

The letters came before Him in reverse order of their emergence. Cosmic *Tav*, the last letter of the *Alef Bet*, came first. She rehearsed carefully the plea she would enter when she presented herself to the Lord. "Lord of the World," she would say, "may it be pleasing in Your Sight to create the world with me, for I conclude Your Seal, which is *Emet*; that is, I am the last letter of the word *Emet* (Truth), which is Your Name.

Because your name is *Emet*, it is impossible for one to cleave to You unless he has acquired the attribute of Truth. It is fitting, then, for the King to start the Creation with the letter *Tav*, which terminates the word *Emet*, for I have the ability to cleave to You."

The *Alef Bet* is not only permeated with the Glory of the Creator, but it is also sealed with the impression of His Signature, which is Truth. To whom is the Lord close? This question is posed in the *Zohar* with regard to the words: "God is near to all those who call upon Him."[87] The *Zohar* clarifies this statement by saying that "God is close to all those who call upon Him in Truth." The *Zohar* then asks: "Is there anyone who would call falsely?" Rav Abba said, "Yes. It is the one who calls and knows not Whom he calls." The *Zohar* raises another question: "What is the meaning of the word *Emet*?" The answer is: "It is the Seal of the King's Ring, which is the Perfection of All."[88]

Tav stepped up to make her case: "Because I stand at the terminating point and do not allow the evil (*klipot*) to draw near and receive nourishment from the Light, there is strong justification for the world to be created with me. For my energy will enable man to distinguish between the Darkness and the Light. Thus will man be assured of attaining his ultimate degree of perfection—the *Gemar HaTikkun* (Final Correction)."

The *Zohar* explains the *Sitra Achra* (the other side), also known as Satan, the negative side, or Dark Lord, obtains its life sustenance from the all-embracing Lightforce referred to as *Kedusha*. The Lightforce keeps the Dark force alive via a minute, precise measurement of energy, so that free will is preserved. Evil, as the student of Kabbalah is aware, is a necessity. For without evil and free will, the universe would revert to its former condition, which allowed no relief from the burden of Bread of Shame.

King David the Psalmist reveals that the Lightforce retains supremacy over Satan: "The Kingdom of the Lord has dominion over all."[89] The unified energy-force of the Light prevails, even in the nethermost places of Darkness. The problem facing mankind is the sustenance we provide to the Dark Lord, although this sharing of sustenance could be terminated at any moment. In those for whom the soul's corrective process (*tikkun*) is complete, there is no longer a necessity for the continued existence of evil (*klipot*), and thus for this person, the sustenance of evil ceases, as does the need for free will.

Tav's sharing of life-giving energy with the Dark force is described in the Book of Proverbs as follows: "Her feet descend unto death."[90] The source of nourishment for the Dark Lord is the "foot" of the letter *Kuf*.

Of all the twenty-two cosmic letters, the letter *Kuf* is the only one to extend below and beyond the line on which the letters rest. This extension of *Kuf* was structured to indicate that a minute quantity of the all-embracing unified Lightforce stretches beneath the permissible cosmic level, indicated by the baseline of the text, into the very depths of galactic consciousness. *Tav* felt that she, too, was capable of extending across the border into the negative realm, but this, as we shall soon learn, was not to be.

The root of falsehood is in the signature of the serpent. The serpent forged the signature of the *Tav* and persuaded Adam to sin by eating from the Tree of Knowledge of Good and Evil.[91] The serpent caused Adam to sin by speaking falsely about the Master of the Universe. The serpent is said to have had legs in the Garden of Eden before Adam and Eve sinned by eating the apple, but by forging the signature of *Tav*, it doomed itself to a life without them.[92] [93] When the Creator learned of the deception, He decreed that the serpent's legs would disappear. Even now, our modern expression of an idea or an endeavor "having no legs" calls to mind the state of impotence that became the eternal fate of the duplicitous serpent. That which has no basis in truth is doomed to go nowhere, so to speak.

The letter *Tav* concludes the word *Mavet* (Death) מות. Thus, *Tav* is said to be the Seal of *Mavet*. Due to the power of *Tav*, death came into existence. Thus, *Tav* is also known as the terminating intelligence. The wisdom of Kabbalah explains that man is subject to death because the serpent forged the signature of *Tav* and made *Adam HaRishon* (the first man) sin by eating from the Tree of Knowledge of Good and Evil.

For two thousand years, the Dark Lord hovered in stationary orbit at the feathered edge of Cosmic Consciousness, receiving a morsel of sustenance once in a while. The leg of the letter *Kuf*, stretching beyond the edge of Cosmic Consciousness, was Satan's lifeline. Satan knew that he could never penetrate the security shield the letter *Tav* maintained around *Kuf*'s leg. But he also knew that the letters were destined to participate in the process of Creation, which meant that one of them eventually would have to descend beneath the line of Universal Consciousness and enter his Dark domain. And so it was that for two millennia, he patiently bided his time, waiting and watching.[94]

Tav was at the outermost position of Universal Consciousness. Her shield protected the entire galaxy, and anything coming in contact with it would be terminated. Cosmic *Malchut* was located at a peripheral area to protect the galaxy of infinity. *Kuf* was even farther out, situated above and preceding *Resh*, *Shin*, and *Tav*. Because *Kuf* was vulnerable to the Dark Lord, her three cosmic companions were necessary for her protection. *Tav* was the perfection of all cosmic letters; she was the Seal of Truth, the protector of Universal Consciousness, her powers impenetrable (or so she thought).

But when *Tav* approached the Lord, her left foot began to be drawn below the line. *Tav*'s desire to be the channel for the process of Creation was all Satan needed. When *Tav* presented herself and said, "Lord of the Universe, may it be pleasing in Your Sight to create the world with me," the Dark Lord, like a deadly scorpion, seized upon her left leg and pulled it into the Darkness.

Satan's empire now would be vastly expanded to encompass the sparks that had been captured after the shattering of the primordial Vessels.[95] The Creator watched in stunned disbelief as *Tav* became enveloped within Dark space. Potentially, she possessed the rigid, negative intelligent energy of judgment so useful in overcoming Satan, but the moment she fell under the influence of the force of the Dark Side, her power was dangerously lessened.

Recognizing the impending disaster, the Creator intervened by curling *Tav*'s left foot upward to make it terminate on a level with the line of Universal Consciousness. Thus, the portion of *Tav* that had been drawn into the Darkness became united once again with the foot itself. The security shield doubled its capacity at the terminating point of the *Tav*, which is why the left foot of *Tav* is thick and wide. And the side of Darkness would have to once again content itself with the morsel provided by the leg of *Kuf*.

When the battle had ended, the Creator spoke to the *Tav* and said, "You are beautiful and upright, *Tav*. However, because you lack the proper attributes, the world cannot be created with you. For just as you are destined to serve as a mark of life upon the foreheads of the faithful who have fulfilled the Bible from *Alef* to *Tav*, so, too, shall your mark designate death for those who fail in their task. Because of your terminating intelligence, one who has not succeeded in fulfilling the Bible from *Alef* to *Tav* will have to perish. For yours is also the Seal of Death."[96]

The Creator continued, "My beloved, faithful *Tav*, you are indeed the Signet of My Kingdom. However, even the righteous must suffer at your behest. Your severe judgment causes them to be punished, not only for their own shortcomings but also for having failed to deter others from their wicked ways.[97] Hence, it would be impossible for the world to survive through you, *Tav*, because of the conclusive and unforgiving nature of your judgments."

The Lord deemed *Tav* unfit to be used for the Creation of the world because she potentially possesses rigid judgments that the world could not endure. Even today, the *Tzadikim Gemurim*

(Totally Righteous), meaning those who have merited the Seal and have fulfilled the Bible from *Alef* to *Tav*, are still punished because of *Tav*'s severe judgments.

Having achieved an understanding of why she would be an unsuitable channel for the process of Creation, *Tav* departed from the presence of the Creator.

CHAPTER 6

THE LETTER
SHIN

ANOTHER GALAXY,
ANOTHER TIME . . .

The original universe—the universe of the Endless World—was far more righteous than any that followed. In that universe, free will had not yet emerged as a viable energy-intelligence, and the prevailing disposition of those who resided within Cosmic Consciousness was such that no intelligent energy within it would so much as consider straying from its predetermined course. Error and indecision were unthinkable, for in the Endless World, *Tav*'s rigid justice reigned supreme.[98] Sometimes she commanded with severity, always with true and accurate judgment. This was the universe the Creator first intended, and He ruled it with an iron will.

At the dawning of the new universe, the emanated Vessels bathed endlessly in the Creator's Infinite Beneficience, but all was not well within the realm of Cosmic Consciousness. The difficulty arose when the Vessels could not, in good conscience, bask in His Light, for they had nothing to give in return. This was the phenomenon mentioned earlier as Bread of Shame. Thus, the Creator found it necessary to create the illusion of a negative space between Himself and His beloved creations. Without it, the Vessels would have no way of relieving Bread of Shame. However, while this illusionary gap served its intended purpose, it also gave rise to negative intelligent energies, beings of Darkness that began to lust for power. At first, under *Tav*'s instantaneous justice, those falling prey to the evil forces were immediately removed from Galactic Consciousness, but more issued forth from the Darkness to replace them.

Having eradicated a sizable percentage of the inhabitants of Cosmic Consciousness for false-heartedness, the Master of the Universe eventually felt compelled to take remedial action. No longer could He remain at a distance. The positive intelligent energies, the guardians of justice in the galaxy, were becoming discouraged as a result of *Tav*'s retribution. No longer could they tolerate the consequences of instantaneous justice. Only one path was available: to remove rigid judgment as the sole form of retaliation for duplicitous thoughts and metaphysical misdeeds. Rigid justice (at least in the primordial sense of that word) could not be allowed to survive, not even within so spiritual a cosmos as Galactic Consciousness. The effect of *Tav*'s swift terminating intelligence would have to be lessened; her sway over that aspect of consciousness would have to end. Henceforth, negative energies were to be treated with *rachamim* (compassion) and given a second chance. No longer would falsehood be instantly condemned. And so it was that a new cosmic intelligence came to power, a letter-energy that would bring an end to *Tav*'s severity and, it was hoped, restore the galaxy to its former pristine glory. The energy-intelligence chosen for this task was cosmic *Shin* שׁ. If all went well, *Malchut* (the physical world) would emerge as a balanced, harmonious entity.

Meanwhile, the Dark Lord, the embodiment of negativity, set out to take control of Cosmic Consciousness. Surrounded by negative intelligences (*klippot*) he set forth to capture every available spark of positive intelligence left over from the Endless World that had passed away with the shattering of the Vessels of Adam. The Dark Lord was determined to establish himself as ruler of the galaxy.

The Creator had other plans. There was only one way to deal with the evil monarch, and that was *not* by fighting fire with fire. The way to destroy Satan was to envelop the Dark aspect of Cosmic Consciousness with the all-embracing Lightforce. This stratagem presented a profound opportunity for *Shin*. Could she bring Light to the Darkness and defeat the evil? Her designation as the Middle Point was certainly significant, but what exactly did it mean?

Concerning this mystery, we refer to the biblical passages: "When you light the lamps…"[99] (Moses said to Aaron), and "Let there be light, and there was light. And the Lord saw the light. It was good."[100] Why was it necessary to repeat the word "light" in the first sentence of the latter verse? The answer is that the first "light" refers to the unified Lightforce symbolized by the Right Column, as exemplified by Aaron, the High Priest. The second "light" refers to the Left Column, which issues from the Right. The words "And the Lord saw the light" refer to the Central Column, *rachamim* (compassion), which unites both the Left and Right sides. And finally, we read: "It was good,"[101] which indicates that the three aspects of the Lightforce—Right, Left, and Central—had been united.

There could be no perfection in the universe until *rachamim* appeared. Only then could the strife that existed between the Right and Left Columns be removed. This is alluded to in the passage: "And the Lord separated between the light and between the darkness."[102] Cosmic *Shin* represents the *Rachamim* Star, the bridge to harmony where the Lightforce could dwell, the embodiment of the Three Columns as portrayed by the shape of her letter, which looks like three columns joined at their base with a straight line. This was the force that would come to the rescue of cosmic *Tav*.[103] Cosmic *Tav*'s rigorous judgment could not overcome the Satan. *Tav* and the Dark Lord had fought to a draw, leaving them both powerless. Then the Master of the Universe intervened, placing

Rachamim, the Star of Compassion, in the Heavens and successfully retrieving *Tav* back into the realm of Cosmic Consciousness.

Space, empty space, and more empty space was the universe that was discovered by the American astronomer Edwin Hubble in the late 1920s. Hubble revealed one of the great mysteries of the cosmos: The universe appears to be expanding. But Hubble's assertion should not necessarily be regarded as the last word. His observations, after all, are speculative, and speculation, by its very nature, is proof of nothing. Thus, we are still free to explore a myriad of possibilities including the seemingly outlandish proposition that perhaps this expanding universe of ours is an illusion—albeit a vital and necessary one—brought about by the process of Creation.[104]

Ever since Copernicus demoted Earth from its central position in the cosmos, it has been difficult to assume that man has a significant place in the universe. The kabbalistic worldview disputes this perspective by declaring that the fate of the cosmos is intimately linked with the fate of Earth and its inhabitants.

The universe has a design, and therefore, a purpose. The *Zohar* tells us that the void of space is an illusion that came into existence when the Lightforce was restricted. This situation might be likened to a lit lantern that has been covered, creating the illusion that its light no longer exists. Cosmic *Shin* was chosen to assist in the removal of this illusion.

To better understand the role that cosmic *Shin* would play in sustaining the universe, we must first examine the three aspects of the Lightforce—the Left, Right, and Central Columns—as they were made manifest in the life of Abraham. The Kabbalah says: "The Lord inspired Abram [later known as Abraham] with a spirit of wisdom so that he was able to discover the essence that maintained balance and the energy-intelligence to which each is entrusted." Genesis states: "And Terach took Abram, his son, and Lot, the son of Charan.... and they went forth with them from Ur of the Chaldees."[105] Because Abram already had left his birthplace, the *Zohar* asks why the Lord repeated the edict to Abram: "Get you forth out from your birthplace."[106] The answer to this question will help us understand the matters before us now.

First, let us examine the power that Nimrod wielded over Abram when he threw Abram into the flaming pit. (The story of the Abraham-versus-Nimrod confrontation appears in the *Midrash Raba*, a major compilation of Jewish scriptural exegesis.) Where did the strength of Nimrod and of the Dark Lord and his negative entities originate? It was born in the negative energy-intelligence known by the code name *Shuruk*, which is the Lightforce that has not been enveloped by the thought-intelligence of sharing (the Light of *Chesed*, or Mercy) and is thus likened to a fire burning out of control.

Nimrod represented the internal intelligence of fire, as is indicated in the verse that reads: "He began to be a mighty one in the Earth."[107] Abram (later Abraham) activated and made manifest the totality of the extraterrestrial energy-intelligences that consist of, and are drawn in the form of, three columns— Left, Right, and Central.[108] When the Left Column energy-intelligence, known by its code name *Shuruk*, made its entrance on the terrestrial level, Nimrod seized this opportunity and used its power to throw Abram—who represents the positive aspect of the Lightforce, the Right Column, and whose energy-intelligence had been the first to appear—into the flaming pit.

We must also understand the secret of Abram's rescue from the flaming pit. When the Lord saw that Abram had been overthrown, He took appropriate action. The only solution was a lightning thrust of Central Column thought-energy that would neutralize the devastating firepower of *Shuruk*. This neutralizing agent, under the code name *Masach DeChirik*, or the Central Column, effectively nullified the Dark Lord's Left Column energy, meaning that the reckless, wanton negative thought-intelligence of Nimrod was no longer able to pour forth chaos and violence within the universe.

This drastic measure restored and maintained a unified balance between the positive and negative poles of thought-intelligence. And now the Lightforce, the Light of Wisdom, became clothed within the spiritual intelligence known by the code name "Light of Mercy." Only when the three aspects of the Lightforce—Left, Right, and Central—are in balance, does the *Mochin*, the embodiment of the Lightforce, become manifest as the grand unification of all universal aspects.

Abram, embodying positive thought-intelligence, was rescued from the flaming pit (the Dark Lord), and it is written that then "they went forth." Thus, the Lightforce again became the all-embracing unified power of the universe. The resulting cosmic euphoria is known by the code name *Yishuv HaOlam*, meaning "the civilized, inhabited world."[109] *Yishuv HaOlam* was the principle according to which the threads of all intelligent energies—all forms of consciousness and their external manifestations—were woven into the fabric of endless, mutually related connections that exist as aspects of the one cosmic whole. With this realization, we come to the end of our search for the ultimate connection between physical reality and consciousness. The cosmic glue—the single unifying energy-intelligence that governs all interactions in the cosmos—is known by the code name *Masach DeChirik*, the Central Column.

Can we really simplify the universe's apparent complexity into a single thought-intelligence? The *Zohar* says we can. The two fundamental and seemingly opposing forces that manifest in innumerable ways, including the apparent attraction and repulsion of the poles of a magnet, are not really distinct forces; they are simply different manifestations of the same underlying interaction that exists in the Endless World.

Before we examine how the energy-intelligence of *Masach DeChirik* restores the two opposite forces into a single cosmic whole, we must define the essential active ingredient of this force. Its name is "restriction," and it is the amazing cure for all of the ills of both the celestial and terrestrial realms.[110]

Can it really be that simple? Are we expected to believe the *Zohar* when it says the cosmos is unified by the all-pervading penetration of the energy-intelligence of man's restriction of his ego? The idea that we can reduce the staggering visible complexity of the universe to its essential simplicity by the power of our own thought-intelligence is an exciting possibility, to say the least. The *Zoharic* saying "as Above, so Below" goes a long way toward describing a universe where all manifestations, both physical and metaphysical, are tied together in a web of interconnected relationships, each separate from, yet joined with, the all-embracing unity.

Because even the most commonplace objects and events are governed by the same laws that regulate the entire universe,[111] it stands to reason that many mundane examples might serve adequately as a starting point for our investigation into the relationship between resistance and the fundamental laws of consciousness. Let us therefore examine the common light bulb. Ever since one of our primordial ancestors discovered that two rocks being struck together produced sparks, man has engaged in a continuous effort to bring artificial light to the hours of darkness. Today, of course, the light bulb is the most common source of artificial light. Yet several fundamental questions remain about this mundane item.

When Thomas Edison set out to invent the incandescent lamp, he was searching for a cleaner, more efficient method of illumination than that provided by gas lanterns or oil lamps. While he obviously succeeded in this endeavor, the final result of his labors was better suited to producing heat than light. In fact, the light bulb is not primarily a lighting system in terms of its energy production. A hundred-watt bulb produces only about five watts of visible light; the other ninety-five watts manifests as heat. In essence, the device we call a "light" bulb is really a tool for transforming electrical energy into low-grade heat energy while producing light as a by-product.

Why does an electric current passed through a filament give rise to so much heat and so little light? Why must there be a positive pole, when, in fact, it is the negative pole that draws the current? Why is a circuit established only after the filament's resistance to the flow of electrical current has taken place? Does a circuit require opposites for its expression? Do opposites truly attract? To answer these questions from the kabbalistic point of view, we must first understand that all events and manifestations are interconnected. The separation of reality into bits, bytes, and quantum fragments is an illusion, kabbalistically speaking, because all manifestations are expressions of a single basic unity. The kabbalists tell us that this illusion was a requirement of Creation and that it is based on the principle of restriction, which the Creator brought into being to remove Bread of Shame and permit free will. The internal force of this illusion is the familiar Desire to Receive for Oneself Alone.

"The universe before the restriction is called 'Infinite' or 'Endless' to denote that in it, there cannot even be imagined an end or limit."[112] Within the Endless World, the Vessel and the Light were different manifestations of the same reality. Any distinction between Vessel (negative energy) and Light (positive energy) is an illusion. Circuitry occurs when the negative aspect, influenced by the energy-intelligence of the filament, is restored to a state of positive intelligent energy.

Opposites do not attract, says the *Zohar*. Space, whether in the galaxies beyond or within the infinite realm of the atom, merely indicates a temporary lack of circuitry. The unlit light bulb reveals the polarities of its components. Once aglow, however, these parts seem to disappear into the light, eliminating whatever space we thought we had previously observed. With the energy-intelligence of restriction at hand, the constituents of matter and the basic phenomena involving them no longer exist as isolated entities.

Einstein alluded to the consolidation of all energy-matter in his proposed Unified Field Theory, which he claimed would one day explain the surface complexity of nature: "Thus all man's perceptions of the world and all his abstract interpretations of reality would merge finally into one, and the deep underlying unity of the universe would be laid bare."

"The energy-intelligence of the Left Column consistently desires to cancel out the illuminating energy-intelligence of the Right Column, which is known by the code name *Cholam*. And since the Right Column energy-intelligence preceded the Left Column in the creative process,[113] its energy-intelligence desires to cancel the strength of negative energy-intelligence and permanently subdue it according to the same principle and relationship which governs the seed, the root, and the branch."[114] The strength of positivity, however, is insufficient to overcome the force of negativity entirely and to subject it to an inclusive relationship with the all-embracing whole.

The task of creating the grand unification of the universe was left to two components of the unified Lightforce. The first, known by the code name *Masach DeChirik*, removes the warhead, or first three *Sefirot*, of negative thought-intelligence. The second, the appearance of an enormous surge of the Light of Mercy, produces the Central Column effect, combining Right and Left into a single unified whole. It is this restriction that allows the negative Left Column to merge with the positive Right Column, thus creating a circuit. Only then is *Malchut*, the World of Action, considered to be part of *Yishuvai Alma*, an orderly, arranged universe.[115]

Let us return to our question of why a light bulb gives off so much heat and so little light. This is caused by the restrictive activity of the filament as it clashes with the negative pole, whose intrinsic intelligence is one of Desire to Receive for Oneself Alone. When the Head (the Upper Three *Sefirot*) is removed, the Lightforce (in this instance, the electrical current) manifests within the bulb at the moment of restriction as a fiery microcosmic birthplace of an illusory universe—yet another of the trillions of miniature Big Bangs that take place every second of every day. As with any explosion, this activity generates a large quantity of heat. At the same time, the energy-intelligence of the Head—the Desire to Receive—is reduced, thus diminishing the full force of the current. Hence, more heat is manifested than light.

Two terminating points are needed for any transfer of energy or, indeed, for any activity of the Lightforce. A third essential constituent in the creation of a circuit of energy is the element of restriction, which is provided by the mediating energy-intelligence known as the Central Column. Cosmic *Tav*—the Vessel of *Malchut*—is devoid of this mediating attribute; within her, *Masach DeChirik*, the Central Column, is nonfunctional. For this reason, the Lord rejected *Tav* as a suitable channel for the Creation of the world. This lack of a mediating energy-intelligence is the reason why a domain of *Din* (severe Judgment/retribution) existed in the Endless, but without this mediating principle, civilization could not exist.[116]

Cosmic *Shin*, however, ends at *Yesod* (Foundation), the level above *Malchut*, where the energy-intelligence of *Masach DeChirik* is functional. For this reason, cosmic *Shin* is also known by the code name *Emet* (Truth) because she embodies the essential unity of all things and events.

Another condition that must be met to assure the harmonious movement of the Lightforce, within which opposites are dynamically unified, is an awareness of the presence of this all-embracing unified intelligence. The code name for the description of this thought-consciousness is *Chotem* (Seal), a word that, as we shall see, embodies various interpretations. How does one achieve this pure awareness? One must understand that this reality, from a kabbalistic point of view, consists of an infinite series of possible worlds, all of which exist within one constant, unchanging Endless World. The fact that our rational minds cannot usually grasp this concept does not rule out the existence of the holistic reality we are describing, the essence of which is infinitely greater than the sum of its parts. Too much evidence in our daily lives, however, suggests that in this physical world, illusion is the rule and not the exception.

The mechanistic view of the universe is totally invalid when applied to altered levels of consciousness. Even the widely held notion that reality is composed of solid objects and empty space is an illusion. In fact, energy and matter are simply aspects of the same space-time/mass/energy continuum, the entirety of which is in constant instantaneous communication. We all have had paranormal experiences, yet we quickly dismiss them because our rational minds lack the ability to comprehend the infinite interconnectedness of our environment.

Chotem (Seal) indicates termination, the end point where energy manifests materially—a symbol aptly likened to the mark of a king at the end of a document. The *Zohar* tells us that the seal of the king commands the same awe and respect as the king's presence. This is because the Signature of the Lightforce (king) personifies cosmic *Shin*, which embodies *Emet* (truth).

"Truth" is another implication of the word *Chotem*.[117] Why was *Chotem* significant as the "Signature of the Lightforce?" The first biblical reference to the code contained within the concept of *Chotem* reads: "And the stones shall be with the names of the children of Israel, twelve, according to their names, like the engravings of the *Chotem*." The *Chotem* was a pouch-like breastpiece worn on the garments of the High Priests of Israel. It was inlaid with twelve precious stones, each engraved with the name of one of the twelve tribes of Israel. Inside a pouch in the *Chotem* were placed the *Urim* and *Tumim*. The priest used the *Urim* to consult the Tetragrammaton for answers.[118] Nachmanides says that the *Urim* was a text bearing the Divine Names, which was placed inside the breastpiece; by virtue of the *Urim*, various letters taken from the names of the tribes became available as channels to higher consciousness.[119] The right to use this oracle was reserved for the Levite priests,[120] and only a king or a head of the Sanhedrin could ask questions of it.[121] King Saul and King David consulted the *Urim*.[122]

The sages have interpreted *Urim* to mean "those whose worlds give Light," and *Tumim* as "those whose worlds are fulfilled." They further explained that the oracle was influenced by rays of Light that shone on the letters, reflecting from them and forming them into groups that the High Priest could interpret.[123]

The *Urim* provided man with a profound insight into the cosmos and his place within it, and thus an unparalleled advance in our understanding of the world. Events could operate not only from past to future, but also from future to past. This was our time machine, our entrance to the Higher Worlds.

Where did the coded information provided by the *Urim* come from? The *Urim* was a connecting chain of intelligence that extended downward from the highest level of universal intelligence by establishing a link between the world of pure Cosmic Consciousness and our species.

With intelligence came power—the power to annihilate as well as the power to maintain an orderly cosmos. This kind of power was fraught with the kinds of dangers that we, even today, know all too well. A powerful person's feelings and values can easily become egotistical and selfish, so there was a risk that a priest or anyone else who knew how to use the *Urim* might fall prey to selfish ambitions and desires. Misuse of this astonishing power was exactly what the Dark Lord had in mind. Such a sophisticated energy system presented the chilling possibility of mankind relinquishing its free will en masse—a condition that would allow the Dark Lord to expand his base of operations, decimate his enemies, and assume complete control over the realm of Cosmic Consciousness.

For the immense power of the Lightforce to ultimately prevail, there would need to be a strong deterrent, a safety valve to ensure that the Death Star fleet could never use the Lightforce. Cosmic *Shin* considered herself worthy of the role of safeguarding harmony in the World to Come.

If we could directly observe the behavior of the atomic world—the ninety-nine percent of existence that is beyond the range of our five senses—we would frequently see objects appearing and disappearing. Past, present, and future would reveal themselves to be unified aspects of the Lightforce. This was the worldview provided by the *Urim*. There was no empty space, no false state of vacuum, no illusion of darkness in the *Urim*—only the all-embracing truth of Light.

Although the metaphysical world seems distinct from the physical world, there is no logical requirement for them to be separated. The unifying power of humanity's conscious restriction is the key to restoring Unity. There is no real boundary between mind and matter. All aspects of physical existence are merely the messengers of consciousness. On a material level, the transference and transformation of energy, matter, and consciousness from one state to another seems hard to imagine until we realize that what distinguishes the animate from the inanimate, energy from matter, is consciousness.

There is no essential difference between the atoms in a rock and the atoms in our bodies. The basic building blocks of both the physical and the metaphysical realms have the same atomic and subatomic structures and tendencies. The difference lies in the four gradations of the Desire to Receive, these gradations being what distinguishes the mineral, plant, animal, and human kingdoms.[124] While many of us believe that only humans have consciousness, the kabbalistic view holds that all four kingdoms are conscious, the chief difference between them being each kingdom's Desire to Receive: The higher life forms have a greater Desire to Receive than the lower ones.

The coded information of the universe—past, present, and future—became manifest within the *Urim*. In these days of computers, the difference between hardware and software is well known. In terms of this distinction, we may conceive of our bodies as the hardware and our internal selves, our consciousness, as the software. For the priests of Israel, their computer was the *Urim*, which represented a program of immense power composed of infinite terabytes of cosmic energy-intelligence.

The essential energy-intelligence of cosmic *Shin* towered above that of cosmic *Tav*. *Shin*, the first letter-energy of the Lord, and is the dominant thought-intelligence of His Lightforce. Cosmic *Shin*'s imposing power is demonstrated by an immensely vital expression of the Lightforce known by the code name *Shaddai* שׁדִּי. The last two letters of *Shaddai*, *Dalet* and *Yud*, together constitute the word *dai*, meaning "enough." Cosmic *Tav*'s negative judgmental intelligence brought about a corresponding expansion of universal energy, which brought with it the danger of overexpansion into unstructured chaos. Thus there was a need for the energy-intelligence of *Shaddai* to enforce a limitation on the proliferation of the Lightforce. All this points to the fact that when the world would later become established as a fitting place for its inhabitants, its completeness would end and manifest at *Yesod* (Foundation) of the World of Action (*Malchut*).[125]

"Enough!" thundered *Shin*. "Expansion must cease! Rigorous judgment must be restrained!" The continuous expansion of the universe would serve no purpose. Cosmic *Shin* subdued the negative intelligence, the cause of cosmic expansion, and brought it into a state of harmonious existence.

And so cosmic *Shin*, armed with *Rachamim* (Compassion), the power to maintain an orderly and well-arranged universe—entered the Lord's presence to state her case. Her powerful position within the Divine Lightforce would provide the power needed to ensure the universe would achieve and maintain proper balance in its manifestation of the Lightforce. Indeed, the code name of the master teacher of Israel, Moses—in Hebrew, Moshe: *Mem*, *Shin*, *Hei* משיה ,—through whom the Redemption of Israel and the world would one day become a reality, embodied the imposing power of *Shin*.

Moreover, the *Shin* is within one of the most powerful of the 72 Names of God: *Mem*, *Hei*, *Shin* שהמ ,[126] the secret code for healing used on *Shabbat.*, and the energy by which the kabbalists could enhance the natural healing potential of all humankind. Through this Name, man can tap the miraculous healing power inside himself. This code enables mankind to harness the power of both thought and emotion by removing any emotional roadblocks that stand in the way of health and happiness.

Cosmic *Shin* was the natural bridge between the world of physical reality and the Lord's internal metaphysical existence of cosmic *Binah* (Understanding).[127] Thus, there seemed to be no logical reason why the Lord should not employ cosmic *Shin* as His channel for Creation.

Shin came forward to propose her thought-intelligence as the suitable channel for the Creation of the universe: "Lord, may it please You to set the genesis in motion through me, as I am the primary thought-intelligence, the first letter of one of the Supreme energy-forces ultimately to manifest at Mount Sinai by way of the cosmic code, the Bible."[128]

And so it was that the Creator's reply to cosmic *Shin* came as a complete surprise to her. "Your attributes, *Shin*," said the Lord, "are indeed most praiseworthy. However, falsehood and deception, the hallmarks and expressions of the Dark Lord, stand no chance of tempting humankind without your active participation. The Master of Evil requires your assistance in establishing free will in the universe. Your energy-intelligence is a necessary component within the agenda of the *klippot*."

Moreover, mankind would not be secure if cosmic *Shin* were to act as the channel for Creation, for no one could be certain to what extent the Dark Lord might go in manipulating her energy-intelligence.[129] The word *sheker* שקר (untruth) consists of three letters: *Shin*, *Kuf*, and *Resh*. If *Shin* were chosen as the channel for Creation, deception and evil might prevail over mankind because *Shin*'s energy is included in the word for falsehood. Without the aid of *Shin*, negativity could not exist, nor could the illusion of free will, without which man would have no opportunity of removing Bread of Shame. Hence, without *Shin*'s active participation within the drama of Universal Consciousness, the very purpose of Creation would be defeated before it even began.

Hearing the Lord's response, cosmic *Shin*, head bent low in humility, departed from the stage of the cosmic process.

Fear of the Unknown;
Cold Infinity;
Planetary Exploration;
Solar System;
Milky Way;
Gravity;
Alternative Universes;
Restriction;
Neutrons;
Force Field;
UFOs;
Quantum Physics;
Extraterrestrial Beings;
Thought-Energy

CHAPTER 7

THE LETTER
RESH

ר

DECEIT AND FOLLY TAKE
COUNSEL TOGETHER AND
CAUSE ME GRIEF OF SOUL.
FALSEHOOD IS THEIR RIGHT
HAND; THEIR MERCHANDISE
IS VIOLENCE, PERJURY,
AND TREACHERY.

—MOSES HAYIM LUZZATO

There are times when we have the feeling that something is out to get us. Fear of the unknown has given rise to many superstitions that have challenged the notion of a Divinely ordered cosmos.[130] Comets, for example, have long been associated with terrestrial catastrophes. Earthquakes, floods, and a host of human disasters, including the fall of kings and presidents, have been attributed to celestial influences. What else but celestial intervention could account for the sudden extinction of plant and animal species that had previously existed for eons?

Mounting scientific evidence indicates that some mass extinction may indeed have been triggered by cosmic activity. One theory suggests that a comet might have collided with the Earth and that the resulting blast so filled the atmosphere with dust that the resulting drop in temperature and the absence of sunlight brought about the mass extinction of many species.

Man has achieved feats of exploration far beyond our wildest imaginings of only a few decades ago. We have set men walking on the moon, and we have sent spacecraft hurtling across the solar system to Mars, Jupiter, and beyond. But such voyages are little more than a trip around the block compared to a voyage across the Milky Way, which would take several million years, even for a spacecraft that could travel at the speed of light. Given our present technology, it appears our best hope for long-range space exploration lies with information carriers such as light or cosmic rays, which may provide a picture of the universe beyond the regions we can explore with spacecraft.

However, because these carriers are extremely limited in their ability to furnish the information we seek, the *Zohar*'s information about the Milky Way takes on a greater importance. The *Zohar* declares that the forces of negativity are located in the Milky Way. It makes reference to the temperature there as too frigid (*kar*) קַר for life-sustaining chemical reactions. The *Zohar* says the Satan's negative intelligences, called *klipot* (shells), are devoid of warmth. Fearing the Light of Wisdom,[131] the very touch of which would obliterate them, these beings of Darkness lurk in cosmic canyons deep in the coldest reaches of the Milky Way, home base of the Dark Lord. Here we find yet another *Zoharic* concept that seems to verge on science fiction, but from the kabbalist's perspective, the evil *klipot* are all too real.[132]

The Dark force is unlike any of the Lord's other creations. Its life-support system is intimately bound up with the letters *Resh* and *Kuf*, which according to the *Zohar* belong to the evil side.[133] *Resh*'s Vessel is the backbone of the Dark force. *Resh*, when combined with *Kuf*, forms the intelligent energy-force of *kar* (cold), the essence of the Dark force. Moreover, the Hebrew word *rash* (poor) רֵישׁ indicates impoverishment, and the *Talmud* declares *Resh* sometimes symbolizes the *rasha* (evildoer).[134] If *Resh* is on the other side (the side of the Dark Lord), how can she also maintain her inseparable link with the Lightforce? Indeed, were it not for the nourishment provided by cosmic *Kuf*, the Dark force would remain in a state of lifeless frigidity, yet while *Resh* and *Kuf* aid in the maintenance of the Dark Lord, both maintain their pure intelligent identity within the Light's twenty-two pure intelligent energy-forces. How can this be? And why?

Perhaps the most interesting feature of both cosmic *Resh* ר and cosmic *Kuf* ק is their similar design: Each cosmic letter has sides rounded at the corners and each lacks a base stroke. Let us pause to examine such a Vessel. At first glance, we observe the absence of an effective collecting area, which means the Vessel is unable to retain a payload of energy. In Kabbalah, projection and retention of energy are crucial components of the workings of the universe. For instance, our universe is energized by the letter *Dalet* ⊤, whose projecting corner indicates that her Vessel is pregnant with the Light of Mercy.[135] One of the functions of the life-force known to kabbalists as the Light of Mercy is to collect and retain the internal energy that kabbalists call the Light of Wisdom. When a Vessel consisting of the Light of Mercy has become unified with the Light of Wisdom, the Earth's gravitational field is weakened. In the kabbalistic view, gravity is created by a specific interaction between the Light of Wisdom and the fluctuating energy field of the Desire to Receive for Oneself Alone.

The propulsion systems of the future will work by altering local gravitational fields. This will be possible when the intelligent energy-force of the Desire to Receive for Oneself Alone has been converted to the intelligent energy-force of the Desire to Receive for the Sake of Imparting.[136] This particular form of transmutation causes electrons, the intelligence of which is primarily the Desire to Receive for Oneself Alone, to reorient their basic internal energy-intelligence. Electrons play a fundamental role in the generation of gravity, so this transmutation causes gravity to weaken.

The Dark Lord and the forces of evil consist of the Desire to Receive for Oneself Alone. The letter *Resh* has only two sides and is rounded at the corner, indicating the lack of a Three-Column System, which in atomic terms is like an atom devoid of a neutron.[137] A circuit of intelligent energy would require the presence of the third, neutral intelligent energy of "restriction," the backbone of the Lightforce and the secret weapon that could ultimately crush the powerful Dark force.

Resh, lacking a restrictive capacity, provides the Dark Side with a perfect vehicle for fulfilling its evil purpose of infiltrating the cosmos with negative or short-circuited energy. However, *Resh* has no desire to embrace the Dark Side; she simply lacks the will to resist it. Like a light bulb with no filament (the restrictive aspect), the shortcoming of *Resh* is merely that she allows Darkness to conquer Light.

The *Zohar* cautions: "Man, by virtue of his evil deeds [succumbing to the Desire to Receive for Oneself Alone], feeds power to the other side." As a result of man's negative acts, the forces of Darkness seize the energy of *Dalet*, a letter that is indicative of the energy-intelligence known to kabbalists as the Light of *Chesed* (Mercy). These negative entities can alter *Dalet*'s consciousness to such an extent that she loses her identity. Her corner softens and disappears, transforming her into a *Resh*. A *Dalet* so captured greatly enhances the power of the Darkness.[138]

The cosmic bond and unifying force of the universe is *Echad* (One) אֶחָד, a most awesome superstructure. The combination of the three letters—*Alef*, *Chet*, and *Dalet*—spelling *Echad* accounts for the basic features of a unified force field. However, whenever the Dark Side captures *Dalet*, this unified field is transformed into the disruptive cosmic influence known as *acher* (other) אַחֵר. When the protruding corner of the *Dalet* (Light of Mercy) is eliminated, a new combination of letters appears in its stead: *acher*—*Alef*, *Chet*, *Resh*. Thus, the prevailing force of the Master of the Universe, which is unity, has been replaced, and strange "other" forces pervade the cosmos, as evidenced by the seemingly endless wars so prevalent in our so-called civilized world.

The crest of the *Kuf* ק's structure is the *Resh* ר. For this reason, the *Kuf* also belongs to the Dark force. *Resh* and *Kuf* accepted a task necessary in the cosmic scheme of humankind to maintain their status within the Light's forces of pure intelligent energy. Without the forces of Darkness, the soul's process of correction (*tikkun*)[139] is disrupted. The removal of Bread of Shame is the fundamental component of freedom of choice. Thus, the Dark force provides humanity with the invaluable—but also undesirable—service of giving us the opportunity, and the duty, of choosing between Light and Dark, good and evil, right and wrong.

The ideas presented here resemble in some ways a science fiction adventure story with a cast of outlandish extraterrestrial intelligences. Indeed, the reader might suspect that instead of a benign tale of the origins of the Hebrew *Alef Bet*, he or she has instead stumbled upon a mad narrative of warfare, espionage, and counterespionage. As a way of understanding our own existence and place in the cosmos, we search for evidence of UFOs and signs of life other than our own, yet extraterrestrials are present in our very midst. If war and hatred appear unavoidable, the blame can be placed directly on the *klipot*. To this day, those evil energy-intelligences maintain a strong hold over many of the inhabitants of our planet. The patriarch Abraham provided us with the tools to detect intelligent signals from other civilizations as well as the knowledge to comprehend the seemingly bizarre nature of life in this and other galaxies. The *Book of Formation* establishes an immutable link between extraterrestrial and earthly life forms, and both the tools and the knowledge are one and the same: the Hebrew *Alef Bet*.

When Sir Isaac Newton formulated his laws of mechanics, many people predicted the end of free will. Newton's theories relegated humanity to a minor position in a mechanistically determined universe. With the advent of quantum mechanics, however, the issue of free will has returned to the forefront of scientific and philosophical inquiry. In one respect, the *Zohar* sides with quantum theory: Both decidedly favor the observer with a vital role in the nature of cosmic reality. According to both the *Zohar* and quantum physics, human consciousness has the unique ability to influence and even radically alter the physical nature of the universe. Heisenberg's Uncertainty Principle points to an inherent indeterminism in the web of the microcosm. Subatomic events have no ultimately definable cause. At the root of this dilemma, from the kabbalistic perspective, lies free will. The subatomic spectrum, which includes man's metaphysical energy-intelligence, is impervious to physical laws. This explains how the energy-intelligence of humankind wreaks havoc in the cosmic flow.

Thus, the struggle between the Lord's twenty-two super-intelligences and the negative intelligences of Darkness depends entirely upon the actions of man. As we mentioned earlier, the *Zohar* states that "Man, by virtue of his evil deeds, accrues power to the Dark Lord." As incredible as it may sound, the fate of our world is influenced by the actions of humankind. Once we are armed with this information, the purpose of the galaxy—and man's role within it—need no longer remain intractable mysteries.

According to the *Zohar*, the forces of Light and Darkness react to the activities of man by reflecting, or projecting, the results of man's behavior back to Earth in the form of comets and bursts of electromagnetic radiation. Red giants and supernovae—those raging, fuming cosmic crucibles—are manifestations of the inhumanity of which man alone is capable. The deaths of stars and solar systems are fueled by man's negative thought-energy, and the shockwaves rippling through space beyond Earth's atmosphere are reflections of man's negative behavior. Yet, while man's power is considerable, our free will is limited because the Lightforce has—and will always maintain—supreme control over the macrocosm of energy-intelligent behavior.

Resh and *Kuf* were given the lonely, dangerous task of maintaining balance within the all-embracing Light of Wisdom that keeps the cosmos from falling into chaos. Thus, *Resh* remains the unsung hero of the Lightforce. Her relative anonymity is indicated by the fact that the Hebrew word *rash*, meaning "poor," found its way into Scripture in the most complex and obscure book of them all: Ecclesiastes. King Solomon, the author of Book of Ecclesiastes, offers the following proverb: "Better is a poor and wise child than an old and foolish king, who knows not how to receive admonition any more. For out of prison, he came forth to be king; although in his kingdom, he was born poor."[140]

The *Zohar* interprets this verse as follows: The "poor and wise child" is the good inclination in a human being—the Desire to

Receive for the Sake of Imparting. This tendency is called "a child" because it is only assigned to a person from the age of thirteen years onward. It is called "poor" because many do not obey it. It is called "wise" because it teaches the right way. The "old and foolish king" is the evil inclination, the Desire to Receive for Oneself Alone. It is called "king" because most obey it. It is called "old" because it is attached to a man from birth through to old age. It is called "foolish" because it teaches the way of evil.[141]

Resh did not make known her desire to be the cosmic channel for Creation, for she knew that she could best serve the Lightforce in other, less obvious ways. Thus, she accepted her destiny, which was to remain an unseen, unsung intelligence working behind the scenes.

Humiliation;
Channels;
Security Shield;
The Lord;
Symbolic Communication;
Manifestation;
Cosmic Forces;
Humankind;
Cosmic Struggle

CHAPTER 8

THE LETTER
KUF

I GAINED RICHES AND
HONOR, AND LIVED BY THEM;
I GAINED IN THEIR TOWNS
AND IN THEIR CASTLES;
THEREFORE THE BIBLE
WAS SLACKED AND THE
EXPOUNDING OF ITS BOOKS
HINDERED.

—DON ISAAC ABARBANEL

osmic *Kuf*, like *Resh*, did not come forward to plead before the Lightforce. Why did she not present herself? Was she ashamed? Had she committed some offense or sin?[142]

Only a fool repeats a blunder, and *Kuf* was no fool. Cosmic *Tav* had made a crucial mistake by revealing her position to the Dark Lord, an error that *Kuf* vowed not to duplicate. Thus it was not shame or humiliation that prevented *Kuf* from presenting herself as a possible channel for Creation; she was simply determined to avoid any unnecessary contact with the *klipot*. Cosmic *Kuf* felt alone in her task, but she did not consider herself in any way inferior to the other letter-energies. On the contrary, she alone had been chosen to venture into Dark space. No other letter-energy was assigned to that dangerous task. It was her mission, and hers alone, to supply the Dark force with the precise amount of energy required to sustain the existence of free will.

The precisely measured ration of energy that reaches Satan is supplied by the leg of *Kuf*.

Cosmic *Kuf* recognized the use made of her ability as a channel for Light, and she understood full well the importance of maintaining the proper balance between Light and Darkness in the universe, for without darkness, free will could not exist. *Kuf* realized, too, that if the process of Creation were to become manifest through her, Satan would be permanently established as a participating intelligent force within Universal Consciousness. Therefore, *Kuf* chased from her mind the very thought of presenting herself as a suitable channel for the creative process, for that thought in itself might have provided Satan with an opportunity to launch an attack.

Why was only *Kuf* capable of withstanding an onslaught from the force of Darkness? Did her inner cosmic strength assure her that she would not be seized by that evil intelligence? Could the security shield provided by *Resh*, *Shin*, and *Tav* protect *Kuf* from being overtaken by the cold Dark Lord? The answer to the mystery surrounding the letter *Kuf* is provided by the *Zohar*.[143] In one of its most mystifying sections—a chapter therefore avoided by most translators and commentators—the *Zohar* explores the methodology of cosmic balance by examining this verse: "And the Lord called unto Moses, and spoke unto him out of the tent."[144]

At first glance, this passage reads as if a conversation between two mortal human beings is taking place. Yet the verse specifically mentions the Lord as the "caller," so how does one reconcile the Lord calling unto Moses? The *Zohar* raises the same question with regard to another passage in Genesis, where we read: "The Lord took Abraham outside."[145] Obviously, the Lord did not take Abraham by the arm and lead him outside, and for this reason, the *Zohar* cautions the reader of the Bible not to be misled by the literal translation. The Bible uses ordinary language to imply cosmic truth beyond the reach of rational consciousness. "Calling unto Moses" and "taking Abraham outside" are figures of speech that indicate metaphysical communication.

Human speech is made manifest by combinations of letters. The Hebrew *Alef Bet* is a conduit for communication, but on a metaphysical level. There is more to etymology than just tracing a word's derivation. An analysis of the position of letters in a word can help us discover information we could not otherwise access. The Hebrew word for "calling," *vayikra*, וַיִּקְרָא mentioned in the verse above, consists of five letters: *Vav, Yud, Kuf, Resh,* and *Alef.* Two of the letters in this word are *Kuf* and *Resh,* which make up *kar* (cold). The Dark force, having no warmth or life of its own, is revealed when the word *kar* becomes manifest. Thus, expressions such as "cold war," "out in the cold," "cold-blooded," "out cold," "cold-hearted," and "in cold blood" all have their roots in the Dark Lord.

A group of scientists at the University of California, Berkeley, holds that the sun is a double star. This theory conforms exactly to the *Zohar's* account of a dual intelligent energy represented by the two triangles of the Star of David—which the *Zohar* calls the sun![146] The Berkeley scientists' theory also corresponds to the *Zoharic* concept of the Dark force. According to the *Zohar,* positive energy-forces are transmitted through the Star of David, while the Dark force is responsible for stormy, negative "weather" as it orbits the plane of the Milky Way.[147] Intelligent negative energy traces its origins to Satan, which accounts for the continuous strife that brings havoc to Earth's inhabitants.

To all appearances, cosmic *Kuf* and cosmic *Resh* belong to the evil side. *Kuf* furnishes sustenance to Satan, and *Resh* assists *Kuf* in their cold war endeavor that pervades our entire universe. However, the *Zohar* cautions that these two cosmic intelligences are also part of the system that directs the awesome power of the Lightforce. Acting as double agents, *Kuf* and *Resh*, along with *Yud* and *Alef*, control the balance of cosmic forces in the universe. This control is indicated by the code word "calling." The role of communication in maintaining the cosmic balance lies at the heart of the verse. At its core, as suggested by their position in the middle of the word, are *Kuf* and *Resh*. *Yud* and *Alef*, the first and fourth letters of the code, protect the perimeters.

Despite the efforts of *Kuf* and *Resh* to serve as double agents, the battle for Universal Consciousness up till now seems to have favored the Dark Lord. In the Age of Aquarius, however, the essence of reality will, according to the *Zohar*, return to its primal state of unification. At that time, humankind will be so engulfed by its own negativity, floundering in a sea of confusion that it will seek to change the cosmos as only it can. With the power of the *Alef Bet* as the instrument of cosmic connection, all manner of detachments and separations will once again be unified, and from the ashes of inadequacy shall arise the possibility of binding everything together once more.

It is said that before the fall of Adam and the subsequent shattering of the Vessels, *Kuf* and *Resh* maintained a perfect unity and symmetry of interactions. Only after the first disruption within the universe did the force of Darkness find an opportunity to manifest negative intelligence. Today our world expresses disunity and lack of symmetry, but we may one day restore this fragmented universe to its original state if we infuse all our relationships with positive intelligent energy.

Cosmic *Kuf* fully recognized the importance of her position within the all-embracing unified Lightforce. Of all the letter-energies, she alone had the energy-intelligence capable of bridging the two parallel universes. Only she could sustain the force of Darkness without yielding to its negative influence.

Were she to leave her post to become the channel of Creation, the danger of the Satan seizing control of Universal Consciousness would be vastly enhanced. With this in mind, cosmic *Kuf* returned to her position, confident of her vital role in the struggle between the forces of good and evil.

Metaphysical Battle;
Outer-space Connection;
Mixture of Good and Evil;
Miracles;
Rav Shimon bar Yochai;
Creation of Man;
Garden of Eden;
Woman;
Life-forms;
Power;
Concealed Light;
The Scroll;
Maturity

CHAPTER 9

THE LETTER
TZADI

ALL MEN'S SOULS ARE
IMMORTAL, BUT THE SOULS
OF THE RIGHTEOUS ARE
IMMORTAL AND DIVINE.

—SOCRATES

Cosmic *Tzadi* saw that *Tav* and *Shin* had been rejected: *Tav* due to the rigid judgments abiding within her, and *Shin* because of the influence placed upon her by the *Sitra Achra* (the evil inclination; lit. the other side). Yet this would not discourage *Tzadi* from venturing forward to propose to the Lord that she might be an effective channel for the act of Creation.

Tzadi, like the letters *Tav* and *Shin*, is included in the word *Chotem*, meaning Signature (of the Lightforce). But unlike with *Shin* and *Tav*, the evil energies of the *klipot* have no access to *Tzadi*. Cosmic *Tzadi*'s immense metaphysical battle station in the Upper Triad of the Star of David (the consciousness of *Zeir Anpin*) affords her ample protection from negative influences. *Zeir Anpin*-consciousness[148] is beyond the range of the Dark forces, which extends no higher than the Lower Triad of the Star of David. Hence, *Tzadi* believed herself to be impervious to the Satan's advances. Her power would be sufficient to easily thwart any attempt by the Dark Lord to seize control of Universal Consciousness, and through her, the world could be guided to its final correction or so she believed.

The cosmic power of *Tav*, *Shin*, *Kuf*, and *Resh* originates within the consciousness of *Malchut*, the level of the greatest concealment of the Lightforce. At this level of consciousness, "the Tree of Knowledge combines both evil and good."[149] Thus, while the intelligent life-forces of the previous letter-energies were admixtures of evil and good (*Malchut* and *Zeir Anpin*), cosmic *Tzadi* drew her energy only from the pure consciousness state of *Zeir Anpin*. Therefore, cosmic *Tzadi* felt qualified to become the channel for the creative process.

At the outset, the letter *Tzadi* looked exactly like the curved letter *Nun* 𝟛 . However, the letter *Yud* of the Tetragrammaton ⸯ surmounted the *Nun* and designated the *Brit* (Holy Covenant) to stand in *Yud*'s stead. The *Brit* is realized through the two parts of the ceremony of circumcision: the cutting of the foreskin and the tearing asunder of the second skin. Compliance with the command of circumcision pushes aside all *klipot*.

Yud, the first letter-energy of *Zeir Anpin*, is the Vessel for the Lord. She is the Emperor of the Tetragrammaton,[150] the most powerful life-force to become manifest within *Zeir Anpin* and *Malchut*. *Yud*, a supreme intelligence, was ready to extend her power as soon as she became manifest. Thus, *Yud* united with *Nun* to create *Tzadi*.

$$\unicode{x05E0} + \unicode{x05D9} = \unicode{x05E6}$$

The Creator's ultimate objective in causing this to happen was to provide sustenance for *Malchut*, the World of Action. *Yud*, the symbol of the *Keter* (Crown) of *Zeir Anpin*, is characterized in the Bible by Adam. Eve symbolizes *Nun*, the female polarity of cosmic *Zeir Anpin*.

Cosmic *Tzadi*, became the most powerful energy complex to exist within *Zeir Anpin*. How and why was such enormous cosmic power achieved within the letter *Tzadi*? Cosmic *Yud* possessed an intense concentration of thought-energy. No other cosmic letter could boast such power. Her bulky yet compact head clearly portrayed the Lord's command station. Cosmic *Yud* enjoyed the supreme metaphysical status of not embodying much corporeal matter. She is, in fact, the smallest letter in the *Alef Bet*.

The Master of the Universe searched for the perfect Vessel to combine with *Yud* to make manifest *Yud*'s extraordinary power. The Lord chose *Nun* as *Yud*'s partner. Cosmic *Nun* permanently faces downward, a graceful sign of her humility, while *Yud* gazes upward. A test in determining whether a person is spiritual or not is to observe the way he or she walks: A head held upward indicates a haughty person, whereas one who walks with head facing downward is generally perceived to be humble. The combination of the positive (*Yud*) and negative (*Nun*) energy-intelligences resulted in the creation of the Lord's supreme Lightforce.

Face to face, *Yud* and *Nun* reigned supreme. *Nun*, formerly in cosmic *Gevurah*, ascended to the Supernal position of *Binah*-consciousness. No energy-intelligence that carried even a trace of the thought-energy of the Desire to Receive for Oneself Alone could possibly survive in the face of this awesome combination. Against a massive onslaught from the *Yud–Nun* combination known as *Tzadi*, the Dark force would disappear with a single burst of Light, leaving no trace of its former existence—or so thought *Tzadi*.

A classic demonstration of this frightening infinite, yet controlled, power is described in the *Talmud*. The legendary sage Rav Shimon bar Yochai was a pupil of Rav Akiva. When Rav Akiva was imprisoned by the Romans for teaching the Bible in public, Rav Shimon continued to study under him and attended to him.[151] Rav Akiva esteemed him greatly, saying to him, "It is sufficient for you that I and your Creator recognize your power."[152] Rav Shimon refused to resign himself to the crushing defeat of the Bar Kochba revolt against the Roman Empire, and he maintained his opposition to the Romans. As a result, Rav Shimon was sentenced to death, and he and his son, Rav Elazar, were forced to flee. For twelve years, they remained hidden in a cave in Peki'in, preserved by a miracle.[153] During this period of solitude, they devoted their complete concentration to the study of Kabbalah and they set their seal upon a view of life that found forceful, eloquent expression in many aggadic dicta.

Elijah the Prophet told Rav Shimon bar Yochai that with the death of the emperor, his own death sentence had been rescinded. After his emergence to freedom, Rav Shimon observed people plowing the fields, whereupon he exclaimed, "How can you leave the realm of spiritual sustenance that provides eternal bliss in exchange for momentary reward?" Rav Shimon and his son commanded such powerful thought-energy that they reduced people to ashes merely by looking at them. Then a voice spoke to Rav Shimon, saying, "Rabbi Shimon, was your exile for the purpose of destroying the world?" The voice ordered Rav Shimon to return to his cave. An additional twelve months in the cave allowed Rav Shimon the time he needed to descend to the earthly plane.[154]

In the World of Action, Rav Shimon's fearsome display of power was not in accord with the cosmic scheme of free will. Man's choice for evil or good cannot and must not be interrupted. The option to choose the illusion of material gain over the all-embracing force of unity has been reserved for humankind from the beginning. Indeed, it was the very purpose of our creation, for without it, we would have no way of relieving Bread of Shame.

When the thought-energy of Desire to Receive for Oneself Alone prevails among the Earth's inhabitants, the forces of Darkness intensify. This is the time when *Nun* must descend and reenter the space of *Zeir Anpin*, falling from her mature position to one of immaturity. This is why a "downfall" is indicated in the letter *Nun*. The Hebrew word for "fall," *nefilah* נְפִילָה, begins with and is imbued by the energy of *Nun*. However, when *Yud* unites with *Nun* to form the letter *Tzadi*, cosmic *Nun* gives rise to the power of cosmic *Yud*. The back-to-back position of the letters *Yud* and *Nun* create this immaturity in that the space between the two letters gives the Dark Lord the opportunity to seize energy. In this state of immaturity, even the combined energy of the two letters has no chance of suppressing the force of Satan.

Throughout the ages, humankind has gained access to one energy source after another, including the energy locked within the heart of the atom. But the creation of pure spiritual energy, such as the energy of our souls, is currently and will always remain beyond our capability to create. In the Bible, we find at least two representations of the act of Divine Creation: "And the Lord created man in His own image, in the image of the Lord created He him; male and female created He them."[155] This verse refers to the creation of man on the Sixth Day of Creation. However, following the Seven Days of Creation, we read another biblical account of man's creation: "Then the Lord formed man of the dust of the ground, and breathed into his nostrils the breath of life; and man became a living soul. And the Lord caused a deep sleep to fall upon the man, and he slept; and He took one of his ribs, and closed up the place with flesh instead thereof. And the rib which the Lord had taken from man, made He a woman, and brought her unto the man."[156]

Is the second account of the creation of man merely a repetition of the first? If so, why was the elaborate description of man's creation not included in the first account? Moreover, if we are to assume that Eve was brought into existence by the process described in the second account, why then was she punished and banished from the Garden of Eden?[157] After all, it had been Adam alone who had been told to refrain from eating the forbidden fruit before the second biblical account of Eve's creation from the rib of Adam. The *Zohar* draws a distinction between the Adam and Eve in the first account of Creation and the Adam and Eve in the second account.[158] "The male and female referred to in the Sixth Day of Creation," states the *Zohar*,[159] "were of a pure intelligent life-thought-form, without any resemblance to corporeal matter." This cosmic state of male and female is analogous to the mature status of *Tzadi* previously mentioned. When *Nun* (Eve) ascends to cosmic *Binah*, the letters turn face-to-face, and maturity prevails. However, when the thought-energy of the Sixth Day was combined with corporeal matter, a back-to-back condition arose. This state of immaturity now exposed Adam and Eve to the clutches of the Satan, embodied as the serpent. Adam and Eve now had the option of free choice: to succumb to the influence of the serpent and eat of the Tree of Knowledge, or to restrict and deny the thought-energy of Desire to Receive for Oneself Alone.

The Divine creation of man had its origins within the formation of the letter *Tzadi*. The state of pure Thought-Energy mentioned in the account of the Sixth Day of Creation was the original face-to-face mature thought of the Master. At that cosmic level, no impure energy (such as that controlled by the Dark force) had the slightest chance of survival. This, then, was the Divine creation of the seven days: the sun and moon, the constellations, fish, water, animals, Adam and Eve—all created as thought-energies. The second account of Creation describes the creation of corporeal matter and the resulting immature back-to-back level of consciousness that gave the Dark Lord the opportunity of seizing pure thought-energies. The chaotic world we live in is a direct result of this seizure, which empowered the Death Star to reign over Earth's inhabitants with terror and destruction.

Woman, symbolized by the letter *Nun*, can either make or break a man, according to the *Talmud*:[160] "A woman is the crown of her husband"[161] and "An evil woman is worse than death."[162] The breaking of a man was the "downfall" indicated by the letter *Nun*. This letter can partake of cosmic maturity or cosmic immaturity—ascending to the heights of pure thought-energy or falling into the abyss of Darkness.

The face-to-face unity and maturity of *Tzadi* was the reason why *Tzadi* thought she would be acceptable to the Lord as a channel for the Creation. Cosmic *Yud*, the male aspect of *Tzadi*, is symbolized on a terrestrial level by the male organ after circumcision.[163] With the removal of the foreskin (*klipot*), the internal energy of *Yud*, now divorced from the admixture of impurities, becomes a dynamic force in the reproduction of future generations. The quality of life encoded in a DNA molecule—the subsequent stage of cosmic *Yud*—depends largely on the degree of *Yud*'s purity. For this reason, the art of circumcision has become synonymous with the *tzadik*, or righteous person. At the same time, the *Sefirotic* power of *Yesod* became associated with the male organ, which underlines the importance of human existence. The development of humankind is directly linked to this awesome ability: the power to create life.

The *Zohar* states: "When we unite in spirituality and share love with our mate at the proper cosmic hour and time, optimum energy becomes manifested,[164] and the union of cosmic male *Zeir Anpin* and female *Malchut* can be consummated." The whole of our cosmos is impregnated with positive intelligent energy. The *Zohar* explores the secret of the covenant of the circumcision and union in its interpretation of the story of Noah, the survivor of the Deluge.[165]

Noah performed a covenant on Earth that corresponded to the metaphysical covenant Above; hence, he is called Man of the Earth.[166] The inner meaning of this covenant is that Noah had need of an ark (feminine polarity, *Malchut*) with which to "unite" so that the seed of all future species would be preserved, as it is written: "to preserve the seed."[167] This physical ark is the Ark of the Covenant, and Noah's union with the ark Below corresponded to a similar union Above of cosmic *Zeir Anpin* and cosmic *Malchut*.

Before the Covenant was established with Noah, he did not enter the ark, as it is written: "...and I will establish My Covenant with you and you shall come into the ark."[168] Noah was a *tzadik*, a righteous man following the celestial pattern. It is written: "The righteous one is the Foundation (*Yesod*) of the world."[169] The righteous one is the pillar that upholds the world. This is implied in the words, "Noah walked with the Lord,"[170] meaning Noah never separated himself from the internal energy-intelligence of cosmic *Yud*. It is also written: "Noah found favor in the eyes of the Lord. Perfect he was in his generation."[171] According to the *Zohar*, the words "perfect he was" indicate that he was born circumcised.[172]

Tzadik, "Righteous," and *Yesod*, "Foundation," are the pillars and protection of the cosmos. Consequently, the letter *Tzadi* thought she would be acceptable to the Lord to be a conduit for His Supreme Energy. Because she was destined to attain maturity and thus perfection, she considered herself to be a fitting cosmic channel for Creation. With this in mind, she entered the presence of the Lightforce.

"Lord of the Universe," *Tzadi* said, "may it please You to create the world through me, inasmuch as I am the sign of the *Tzadikim* (Righteous) and of Yourself, Who is also called Righteous. For as it is written: 'The Lord is righteous, He loves righteousness.'"

The Lord answered, "Yes, *Tzadi*, you signify righteousness, but righteousness must not be needlessly exposed. Were the Creation of the world to begin with *Tzadi*, you would become unveiled and give the world cause for offense. The Dark force must be part of the universal landscape to provide corporeal man with free will. Your disclosure could lead to Satan's seizure of the power of cosmic *Yud*. You must, therefore, remain in a back-to-back configuration, which designates that the Light is concealed."

The Bible Scroll presented on Mount Sinai gives evidence of the Master's refusal of cosmic *Tzadi*. The letter *Tzadi* found in the Bible Scroll was positioned so that cosmic *Yud* of the *Tzadi* faced outward and away from her companion, *Nun*, indicating that the universe was not yet ready for cosmic maturity.

In consoling *Tzadi*, the Master of the Universe assured her that in time she would ascend to a higher realm of cosmic consciousness, a level at which the *Yud* aspect would reign supreme. Secure in the knowledge that she would one day be restored to face-to-face maturity, *Tzadi* returned to her place in the cosmic panoply.

Freedom;
Law of Motion;
Space;
As Above, so Below;
Israel;
Empires;
Mount Zion;
Holy Temple;
Cosmic Darkness;
Astrological Charts;
Incarnations;
Earthquakes;
Physical Reality

CHAPTER 10

THE LETTER
PEI

I CALL HEAVEN AND EARTH
TO WITNESS AGAINST YOU
THIS DAY THAT I HAVE SET
BEFORE YOU LIFE AND
DEATH, THE BLESSING AND
THE CURSE; THEREFORE
CHOOSE LIFE.

—DEUTERONOMY 30:19

The letter *Pei* came forward on the stage of Creation and pleaded, "May it please You, Master of the Universe, to create through me the cosmos. I signify *Purkanah*, emancipation. Your design for the intelligent energy-force to provide emancipation for the entire world is the letter *Pei*. Freedom for the entire world becomes manifest through a combination of letters forming the cosmic word *Pedut* (Redemption). The source and first letter of this power force is the letter *Pei*. Consequently, it is fitting that the Creation of the universe be channeled through me."[173]

Conforming to the kabbalistic dictum "as Above, so Below," we expect to find the positive influence alongside the negative. Wherever intense positive energy is found, the forces of evil lie in wait for the opportunity to seize it. "Sin crouches at the door; and unto you is its desire, but you may rule over it," declares the Bible.[174] Sir Isaac Newton expressed this important universal law of nature when he stated his Third Law of Motion: For every action, there is an equal and opposite reaction. Kabbalists call this concept the Law of Two Systems: [175] "The Lord has set the one opposed to the other," proclaimed King Solomon.[176]

As we study Kabbalah, we notice how the energy embodied in one manifestation permeates and prevails in all the other kingdoms of Creation, each being but a different expression of the all-embracing unified reality.

The efficacy of vaccination, for example, serves to verify this fundamental rule of universal interaction: The polio virus prevents polio, and snake venom provides the best way to combat a snake bite. We see this interaction in nature, too, as alongside the poison ivy branch, we discover jewel weed, its antidote.

The intelligent energy-force, flowing by means of the Aramaic word *Purkanah*, is also referred to by the arrangement of the Hebrew letters structuring the word *Geulah* גְּאוּלָה, meaning Redemption. *Geulah*, also indicating freedom, is closely aligned with the Hebrew word *Galut* גָּלוּת, meaning Exile. "The glory of Israel has gone into exile, for the Ark of the Lord was taken away," proclaims the Scripture.[177] The cosmic letter-energy that differentiates between the opposing concepts of exile and freedom is the letter *Alef*. The cosmic structure of *Geulah* (Redemption) includes the *Alef*, whereas *Galut* (Exile) lacks the *Alef*, which is the letter-energy that brings about the emancipation of energy-intelligence throughout the entire galaxy.

Large-scale physical phenomena are conditioned by the microscopic organization of intelligent forces. The progress of entropy is controlled by the motion of molecules. The macro material world is completely supported by the micro world; each is a reflection of the other. War and peace, exile and freedom are reflections of the broader metaphysical universe. The universe mirrors a pattern of human existence consistent with the basic laws of both the celestial and the terrestrial realms. There is as much rivalry in the Heavens as there is on Earth. Space is a battleground. A fierce struggle continually rages between the cosmic forces. The universe, which burst into existence as a result of a savage explosion, still brims with violent activity—eruptions, convulsions, and matter lost in the oblivion of black holes. All across the vast cosmic wilderness, the eternal struggle continues.

Exile and freedom, then, are merely reflections of the larger universe. But what triggers them? Why do freedom or exile come and go at precise times? Freedom and exile are dependent upon the status of *Malchut*, the feminine polarity. Whenever *Malchut* lacks the *Mochin* (the internal force of the brain), which embodies the Upper Three *Sefirot—Keter, Chochmah,* and *Binah*—the Israelites are driven out of the land of Israel and into exile. This phenomenon is simply the specific expression of the natural law "as Above, so Below."[178]

The land of Israel is the energy center of our physical world, and the land of Israel Below is affected by the land of Israel Above. The Israelites are the physical expression of the cosmic *Malchut*, while the land of Israel is the male polarity of *Zeir Anpin*. When cosmic male (*Zeir Anpin*) and cosmic female (*Malchut*) are in harmony, a complete energy circuit is established, and at such times, the Israelites are connected to the land of Israel. But when Israelites Below create a short circuit with negative deeds, a schism is established between *Zeir Anpin*, the positive polarity, and *Malchut*, the negative polarity, alienating one from the other. This short circuit generates violence in the celestial realm.

Who are these people called Israelites? Why have they stamped such an extraordinary impression on the universal stage of history, out of all proportion to their numbers? "Israelites maintain the most intense degree of the Desire to Receive of all other nations of the world," states the *Zohar*.[179] This level of negativity provides the opportunity for maximum expression of the all-embracing Light of Wisdom. Just as a light bulb draws energy according to the restrictive capacity of its filament, the degree of Light manifested by humankind depends upon its ability to restrict and thus draw that sublime cosmic energy. Restrictive ability reaches its highest manifestation in the Israelite, and the region through which the maximum energy reaches our galaxy is the land of Israel. "When the Lord created the world, He created seven firmaments Above. In each one dwell stars, constellations, and ministers to serve. Similarly, there are seven spiritual divisions on Earth Below, one spiritually higher in energy than the other, the land of Israel being the highest of all, and Jerusalem being the energy center of the whole inhabited world."[180]

The universe is a message written in a cosmic code, the key to which is contained in the *Zohar*. When we decipher the code, we behold a celestial order beyond our rational comprehension. When the Scriptures refer to the region of "the land of Israel," they are talking about this invisible realm of consciousness. When pondering the *Zohar*'s description of the land of Israel, it is fascinating to examine the relationship between Israel and the rise and fall of former empires.[181] The Parthian Empire of Cyrus, for example, reached its peak when Cyrus overtook the land of Israel. A similar destiny awaited the Babylonian, Greek, and Roman empires, and more recently, the Ottoman Empire. The most contemporary example of this extraordinary pattern is the British Empire. Astonishingly, each empire reached the peak of its international influence when it ruled over Israel, and each measured its decline from the time it lost possession of the land of Israel.

The intrinsic internal dynamics of the land of Israel exactly parallel the intensity of the Desire to Receive manifested by Israel's inhabitants. The particular cosmic intelligence generated by this land seeks a corresponding level of receptivity because positive energy wants to share. The people of Israel sustain the power of the land by generating the highest form of the Desire to Receive, which is the Vessel for revealing the Light of Wisdom. The capacity of Israel's Vessel corresponds precisely to the amount and intensity of energy flowing forth from the land. For this reason, the region has been known as the land of the people of Israel—a code for the interconnectedness of the people and the land of Israel.

Now we will examine how the land of Israel became known as the Holy Land. Following the destruction of Shiloh (ca. 1050 BCE), Israel needed a central Temple. The military defeat suffered by the Israelites at Eben-Ezer, which ended with the Philistines capturing the Ark of the Covenant, brought about a severance of the Ark of the Covenant from the altar. When King David took Jerusalem as his capital, he brought the Ark of the Covenant back to Mount Zion, where he erected a tent for its safekeeping.[182]

Scholars are divided in their opinions as to why David chose that specific mountain in Jerusalem as the resting place for the Ark. The main question was whether the spot selected for the altar was also the place that tradition had identified as the site of the binding of Isaac.[183] But another factor in choosing Jerusalem as a site was its territorial independence. As a newly conquered city, it had not yet been incorporated into the territory of any one tribe. Jerusalem was thus the only site likely to satisfy every tribe as a repository for the Ark.[184]

It is true that Abraham bound his son Isaac on this site, and Jerusalem's territorial independence is a matter of historical record. Yet many questions still remain unanswered. Why was this particular site chosen for Isaac's binding? Why did Jerusalem remain territorially independent? Why is Jerusalem designated a Holy City? Was it simply because the Holy Temple was located within it?

The answers to these and a host of other questions lie in the energy that is inherent in the land of Israel. The *Zohar* tells us that both the Temple and the Ark were collectors and conductors of cosmic energy-intelligence. When a circuit of energy flowed, the universe and all of its infinite galaxies were in harmony and violence did not exist. "When Israel dwelt in the Holy Land, the Lord sent down food to them from a Supernal region, the surplus of which was given to the rest of the world."[185] But when the children of Israel sinned, a void was created between the land of Israel Above (*Zeir Anpin*) and the people of Israel Below (*Malchut*). Consequently, the people of Israel were driven out of the land of Israel, indicating a lack of connection to *Mochin* (elevated spiritual Light), and the world and the cosmos were thrust into short-circuitry, darkness, and violence. When the Israelites corrected their misdeeds, they caused Israel Above to bestow Light upon Israel Below; the land of Israel and its people were again reunited, and Light illuminated the cosmic Darkness. This joyful condition was dramatized by Israel's return as a spiritual people to its land.[186]

The mystical significance of the cosmos was well known to all ancient cultures. Temples, monuments, and religious teachings stand as testimonials to the worldwide belief that the Heavens have always exercised influence over the daily lives of humankind. The ordered structure of the cosmos was thought to symbolize the metaphysical workings of the Celestial Realm. The Temple of Jerusalem, however, held a different status than did the temples of other ancient peoples. What happened in the Temple—and therefore in Jerusalem—was thought to affect everything both on Earth and in the cosmos. The ancient Israelites believed that Earth, rather than being just one among countless billions of celestial entities scattered throughout the universe, was the center of the universe, and Jerusalem was considered to be the nucleus around which all the galaxies revolved. The Temple was thought to give nourishment, peace, and prosperity to all inhabitants of the Earth and even to extraterrestrial intelligences throughout the cosmos. The Ark of the Covenant was the instrument that drew the Supreme energy-intelligence down to Earth. The Israelites considered man to be the prime interface between terrestrial energies and the celestial forces of the metaphysical domain, and only when the forces were in balance did the Heavens declare their majesty and influence.

Cosmic *Pei* symbolizes the channel by which freedom would reign throughout the cosmos. Man's good deeds produce a harvest of positive intelligent energy throughout the Celestial Realm. Cosmic *Pei* acts as the transferring channel by which the cosmos ultimately beams this manifested positive energy throughout the galaxies, decreeing the end of violence in both the terrestrial and the celestial dominions. And because the cosmic order was assured by the intelligent energy-force of freedom provided by *Pei*, she felt especially qualified to act as a channel for the Creation of the world.

For thousands of years, people believed the universe was like a marionette whose strings were manipulated by celestial entities. Then, some 3800 years ago, the patriarch Abraham, the world's first known astronomer/astrologer, revealed that the powerful celestial bodies in the sky were not gods, but rather, energy-intelligences whose sole function was to execute and manifest the activities of the Earth's inhabitants.

In many beautiful legends, the *Midrash* recounts how Abraham turned away from the debasing heathenism of his contemporaries.[187] One night in his early childhood, Abraham gazed at the stars and thought, "These must be the gods." However, at dawn the stars disappeared, and when the sun rose, he exclaimed, "This is my god: him will I adore." Then the sun set and he hailed the moon as his deity. But when the moon, too, disappeared, he cried out, "This is no god! There must be one Creator of the sun, the moon, and the stars."

In Abraham's father's house, there stood one great idol and a number of smaller ones. Abraham broke all of the smaller idols and placed the hammer in the hand of the larger one. "They quarreled amongst themselves," explained Abraham to his confused father, "and the big one took a hammer and smashed them all. See, the hammer is still in his hands."

"But there is no life or energy in them to do such things," answered his father.

"Then why do you serve them?" asked Abraham.

And so it came to be that celestial entities would no longer be considered deities; instead, they would be accurately identified as intelligent cosmic channels. This concept is reinforced by the *Zohar*, which states that the behavior of the stars and planets is not causal, as was previously imagined. Rather, the celestial entities act as mirrors, reflecting events taking place on Earth.

"The stars impel, but they do not compel," declares an ancient kabbalistic expression. An event and its astrological portrayal are not necessarily linked. Astrology focuses the space-time continuum onto a horoscope. An astrological map can answer a given question, but free will gives an individual the power to choose an alternate course.

Kabbalah teaches that the implications of one's astrological chart are based upon prior incarnations. Each of us has a movie of our life. This teaching might be likened to a movie from the 1970s that can be remade in the twenty-first century. While the film was originally made at an earlier time, this time around, it is your film and you can do with it as you wish. In the case of your current incarnation, you are the producer, director, star, and distributor, and as such, you have the artistic license to edit, re-shoot, cut, dub, or add material at will. Each new release gives you another opportunity to complete your soul's corrective process and achieve your own unique *tikkun*.

The letter-energy *Pei* is our meditative channel for altering the future events of our lives. While our stars impel us based on the energies expressed during prior incarnations, *Pei* gives us the power to redirect our energies to higher, more productive states of consciousness. Human activity can change the outcome of the remade film for better or for worse.

The astrological chart is a signpost. It can remove us momentarily from the raging storm of material experience and guide us toward the more subtle aspects of our lives. Astrological meditations are channels we can use to achieve altered states of consciousness and thus arrive at the spiritual realms where metaphysical changes can occur. Human intervention in these realms can disrupt the flow of time, preventing events that seem inevitable. The ability to predict the future can also arise from such meditations. The process of editing and re-editing knows no bounds. Yesterday, today, tomorrow—all are one and the same, once we have stepped beyond the limits of rational consciousness.

Rav Shimon bar Yochai, the master Kabbalist, is famous for altering the direction of time and events. The Light that emanated from Rav Shimon bar Yochai was said to be enormously intense. Such a Light, said the sages, will illuminate all Creation at the end of the period of correction. The *Zohar* illustrates Rav Shimon's ability with the following story:

One day, Rav Shimon observed that the world was covered by Darkness and the Light was concealed. His son, Elazar, said to him, "Let us try to find out what the Creator means to accomplish." An angel appeared to them in the form of a great mountain, spewing forth thirty torches of fire. Rav Shimon asked the angel what he intended to do. "I am instructed to destroy the world," said the angel, "because mankind does not contain in its midst thirty righteous individuals." Rav Shimon replied, "Go before the Creator and tell him that bar Yochai is among the inhabitants of the world. My merit is equal to that of thirty righteous men."

The angel ascended to the Creator and said, "Lord of the Universe, are you aware of bar Yochai's words to me?" The Creator replied, "Descend and destroy the world as you were commanded. Take no notice of bar Yochai."

Seeing the angel reappear, Rav Shimon told him, "If you do not ascend again to the Lord with my request, I shall prevent you from ever reaching the Heavens again. And this time, tell Him that if the world lacks thirty righteous men, He should spare it for the sake of ten. If there are not ten such men to be found in the whole world, then ask him to spare it for the sake of two men, my son and me. And if He deems these two insufficient, then ask him to preserve the world for the sake of one man, and I am that one. For the Scripture states: 'But the righteous is an everlasting foundation.'"[188]

At that moment, a voice from Heaven said, "Praiseworthy is your portion, Shimon bar Yochai, for the Lord issues a decree, and you seek to nullify it; surely for you the psalmist wrote the verse:[189] 'He will fulfill the desire of those that fear Him.'"[190]

Here we see demonstrated the power of altered states of consciousness. Rav Shimon, having acquired the highest level of both Inner and Encircling Light, challenged the authority of the cosmos and succeeded in altering its direction.[191] This, then, was the power of *Pei*. Her cosmic capability would provide Earth's inhabitants with the possibility of thwarting the great tribulations facing them. Can the future violence that threatens our planet be halted? The answer was given by the Master of the Universe when he spoke to Rav Shimon bar Yochai. Good deeds by the Earth's inhabitants enhance *Pei*'s cosmic ability to order terrestrial and extraterrestrial activities. People of extraordinary spiritual ability can change the destiny of the cosmos.

Perhaps the most fantastic story of the extraordinary influence of mind over matter concerns Joshua ben *Nun*. When pursuing the Canaanite kings at Beth-Horon, he implored the sun and the moon to stand still. "And he said in the presence of Israel, 'Sun, stand still upon Gibeon; Moon, in the valley of Ayalon.' So the sun stood still in the midst of Heaven and did not go down about a whole day."[192]

Until quite recently, that biblical narrative was regarded as impossible. Jericho's walls might possibly have been breached by an earthquake;[193] the parting of the Red Sea could perhaps be explained by a freak tornado[194]—but what natural catastrophe could have halted the Earth's rotation? Due to the rise in popularity of such radical theories as those expressed by author and scholar Immanuel Velikovsky, who reinterprets events from history by reconciling biblical history and modern archeology, a departure of the Earth from its regular axis is acknowledged to at least be a possibility. Now it is commonly believed that at times during its history, the Earth may have passed near enough to a heavenly body of sufficient mass to disrupt Earth's orbit. A comet or meteorite may also have struck the planet, causing it to temporarily stop following its normal orbit.

But while this biblical narrative now finds a degree of scientific support, many others do not. For instance, the possibility that consciousness might play a role in celestial activity remains unaccepted by all but a few radical metaphysical thinkers. The vast majority of traditional scientists, clergy, and laymen would not entertain for a moment the idea that the human mind could influence remote physical phenomena on a grand scale. Yet the Scripture asks us to accept without question that consciousness can enable certain individuals to transcend the laws of physics. Who, or what, are we to believe?

A new recognition of the role of consciousness in the universe is slowly but surely emerging within the new physics. Quantum theory affirms scriptural claims by suggesting that human intervention influences the structure of physical reality. This "new" quantum perspective asserts a radical departure from the mechanistic Newtonian view of reality—and a return to kabbalistic values.

How did Moses part the Red Sea? Few would deny that a phenomenon of great significance took place in Egypt that day. However, as with the story of Joshua ben *Nun*, some scholars have attempted to mitigate the importance of the Red Sea phenomenon, relegating it to the realm of myth and legend. Others have assumed that freak tides or other climatic phenomena were responsible. But even if those theories are true, they do nothing to explain why these extraordinary events occurred where and when they did.

The commentator on the passage in Scripture says, "The water of all oceans and seas was divided."[195] A description of the parting of the Red Sea provided in the *Midrash* says: "The waters were piled up to the height of sixteen hundred miles, and they could be seen by all the nations of the Earth."[196] Whatever the circumstances, this event must surely be ranked among the most dramatic milestones in the annals of Jewish, if not human, history. Why should the Israelites have escaped destruction while their oppressors perished before their very eyes?

"And the pillar of the cloud removed from before them and stood behind them."[197] What was this pillar of cloud? Rav Yosi, in the *Zohar* portion of Beshalach 9:157, hypothesized that it was the cloud that is always seen with the *Shechinah*,[198] the cloud into which Moses entered.[199] The wisdom of Kabbalah describes and explains the power of Moses thus: From the side of *Chesed*, the Right Column, there are seventy-two letters; from the side of *Gevurah*, the Left Column, there are seventy-two letters; and from *Tiferet*, the Central Column, there are seventy-two letters. In the transcendent sphere, the letter-energies are all linked together, forming the Divine Chariot, the Holy Name. In these three verses, which spell out the seventy-two letters of the Holy Name, are inscribed the three elements of water, fire, and air."[200]

First Verse Right Column

וַיִּסַּע מַלְאַךְ הָאֱלֹהִים הַהֹלֵךְ לִפְנֵי מַחֲנֵה יִשְׂרָאֵל
וַיֵּלֶךְ מֵאַחֲרֵיהֶם וַיִּסַּע עַמּוּד הֶעָנָן מִפְּנֵיהֶם וַיַּעֲמֹד
מֵאַחֲרֵיהֶם׃

8	7	6	5	4	3	2	1
מ	מ	ע	א	שׁ	ל	א	ו
ד	פ	ע	ח	ר	פ	כ	י
מ	ג	מ	ר	א	ג	ה	ס
א	י	ו	י	ל	י	י	ע
ח	ה	ד	ה	ו	מ	ם	מ
ר	ם	ה	ם	י	ח	ה	ל
י	ו	ע	ו	ל	ג	ה	א
ה	י	ג	י	ר	ה	ל	ר
ם	ע	ן	ס	מ	י	ר	ה

Scanning directions

Second Verse Left Column

וַיָּבֹא בֵּין מַחֲנֵה מִצְרַיִם וּבֵין מַחֲנֵה יִשְׂרָאֵל וַיְהִי
הֶעָנָן וְהַחֹשֶׁךְ וַיָּאֶר אֶת הַלַּיְלָה וְלֹא קָרַב זֶה אֶל זֶה
כָּל הַלָּיְלָה:

ה	כ	ל	ה	ל	י	ל	ה	9
ק	ר	ב	ז	ה	א	ל	ז	8
ה	ל	י	כ	ה	ו	ל	א	7
ש	ר	ו	י	א	ר	א	ת	6
י	ה	ע	נ	ן	ו	ה	ו	5
י	ש	ר	א	ל	ו	י	ה	4
ו	ב	י	ן	מ	ו	נ	ה	3
ו	נ	ה	מ	צ	ר	י	ם	2
ו	נ	ה	מ	צ	ר	י	ם	1
ו	י	ב	א	ב	י	ן	מ	0

→ Scanning directions

Third Verse Central Column

וַיֵּט מֹשֶׁה אֶת יָדוֹ עַל הַיָּם וַיּוֹלֶךְ יְהוָה אֶת הַיָּם בְּרוּחַ
קָדִים עַזָּה כָּל הַלַּיְלָה וַיָּשֶׂם אֶת הַיָּם לֶחָרָבָה
וַיִּבָּקְעוּ הַמָּיִם:

8	7	6	5	4	3	2	1
י	ה	י	י	ה	ו	ד	ו
ב	י	ל	ם	י	ל	ו	י
ק	ם	ה	ע	ם	ר	ע	ט
ע	ל	ו	ז	ב	י	ל	מ
ו	ו	י	ה	ר	ה	ה	ש
ה	ר	ש	כ	ו	ו	י	ה
מ	ב	ם	ל	ו	ה	ם	א
י	ה	א	ה	ק	א	ו	ת
ם	ו	ת	ל	ד	ת	י	י

Scanning directions ↓

The 72 Names of God

Scanning directions →

8	7	6	5	4	3	2	1
כהת	אכא	ללה	מהש	עלם	סיט	ילי	והו
הקם	הרי	מבה	יזל	ההע	לאו	אלד	הזי
והו	מלה	ייי	נלך	פהל	לוו	כלי	לאו
ועיר	לכב	אום	ריי	עאה	ירת	האא	נתה
ייז	רהע	וזעם	אני	מנד	כוק	להו	יוז
בייה	עעיל	ערי	סאל	ילה	וול	מיכ	ההה
פוי	מבה	נית	נגא	עמם	הועי	דני	והו
מוב	עני	יהה	ומב	מצר	הרוו	ייל	נמם
מום	היי	יבם	ראה	ובו	איע	מנק	דמב

Scanning directions ↓

"And Moses said to Joshua…"[201] Joshua was but a youth, and Israel held many warriors greater than he. Why, then, did Moses speak to him?[202] The reason was that Moses, in his wisdom, knew that what was about to happen was not going to be merely a battle of flesh against flesh—Israelite against Amalekite—but a contest pitting the energy-intelligence of good against the extraterrestrial intelligence of evil. Joshua, though a youth, had already reached an elevated state of spiritual consciousness. While not as high as Moses' soul, which was united with the *Shechinah*, Joshua's soul had attained the extraterrestrial region of intelligence called *Metatron*, meaning "youth."

When Moses perceived that the Dark Lord, Samael, was going to assist the nation of Amalek, Moses thought: "This young man, Joshua, will surely stand against the Dark Lord and prevail." He therefore said to Joshua, "Go and fight against

Amalek. It is your battle, the battle here Below, and I will prepare myself for the battle Above."[203]

This account concurs with the kabbalistic description of the seventy-two letters. The knowledge of this devastating power was transmitted via the letter *Pei*, who provided the full impact of freedom to the Upper and Lower cosmos. It now becomes apparent why the letter *Pei*, the cosmic channel for the *Sefira* of *Hod-* (Glory) consciousness of *Zeir Anpin*, felt worthy of presenting her case to the Master of the Universe.

With this in mind, *Pei* entered the presence of the Lightforce to state her plea. "*Purkanah* (deliverance), which You shall ultimately bring to pass in the world, is indicated within cosmic *Pei*," she said. "The all-embracing, unifying bond is eternally clothed within me. The *Mochin* of *Malchut* contains the Lightforce for the redemption of the galaxies. When this Light becomes cosmically connected with *Malchut*, the feminine polarity will manifest as freedom from all forms of violence. Thus, only through me can the entire universe become enhanced to the state of its Final Correction."[204]

What is the cosmic force that stands in the way of universal freedom? The Final Redemption of *Malchut* from the clutches of the Dark Lord is closely bound up with *Zeir Anpin*'s ability to connect with the *Mochin* from *Ima* (*Binah*). Everything depends upon *Malchut*'s level of purification from the *dinim* (judgments). The *dinim*, states the *Zohar*, were intelligent life-forces that came about as a result of the Restriction.[205] This influence placed a limitation upon the universal bonding force of the all-embracing unified reality. Freedom and the Final Correction, then, are expressed when *Ima* (*Binah*), the Supernal Mother of

Zeir Anpin and *Malchut*, releases and provides her daughter, *Malchut*, with the channel to remove the *dinim*.

Without proper communication with *Binah*, *Malchut* is subject to the cosmic influence of the *dinim*, and the channels of mercy and compassion are prevented from making their influence felt in the cosmos. The *dinim* obstruct the release and emission of these two intelligent life-forces that provide the fundamental harmony in the universe.

The process of obtaining spiritual freedom is described in the *Midrash*, where the sages point to the eagle as the symbol of cosmic *Binah*. The eagle's compassion for her children is expressed when she places them upon her wings to protect them from the enemy's arrow. She says, "I prefer to have the arrow strike me rather than my children." This profound mercy, manifested in *Malchut*, is also revealed in the Bible: "As an eagle stirs up her nest, flutters over her young, spreads abroad her wings, takes them, and bears them on her wings."[206]

"You are indeed very beautiful," the Master of the Universe said to *Pei*. "But within you, *Pei*—penetrating the inner recesses of your cosmic intelligence—lies the impression of *pesh'a*. Your very essence is the first letter of the word 'crime.' Thus, while you embody the intelligent life force of *Pedut*, Redemption, the *Mochin* that springs forth from *Ima* cannot be permanently established within your cosmic sphere of influence."

To more fully understand this reply from the Master of the Universe, we must discuss the energy that brought forth the original man, Adam. When Adam was born, he possessed the level of consciousness designated as cosmic *Kedusha*, the outer-space connection,[207] which is located within the consciousness of *Zeir Anpin*. Adam's scope and dimension extended beyond the vast expanse of our galaxy, sweeping across all of Cosmic Consciousness and the World (*Asiyah*, or corporeal action), and beyond infinity to the World of *Briah* consciousness.

The Tree of Life colony mentioned in the Bible[208] connected and communicated with the realm of the worlds of cosmic *Kedusha*. The Tree of Knowledge of Good and Evil, on the other side of Cosmic Consciousness, was located in the worlds of cosmic *Tumah*, the realm of Satan. Adam was commanded not to eat from the Tree of Knowledge.[209] By maintaining this cosmic distance, the Master of the Universe kept his Kingdom safe and secure, beyond the far-reaching tentacles of the Dark Lord. There was, under no circumstances, to be contact with the *klipot*.

However, Adam sinned, thereby establishing contact with the *klipot*, and he was drawn, as if by a magnet, into the Dark Lord's domain. Along with Adam, the worlds of cosmic *Kedusha* entered the three contaminated worlds of the Dark Lord. Death, violence, imperfection, and immaturity now began to rule supreme, and this catastrophic state of imbalance will continue until Messiah-consciousness is universally revealed.

Let us reflect for a moment on the standard interpretation of the biblical account of Adam's sin. Eating the forbidden fruit created contact with the other side. The fruits of both trees embodied the entirety of each consciousness: *Kedusha* (Holiness, Purity) for the Tree of Life and *Tumah* (Impurity) for the Tree of Knowledge, a situation not unlike our modern understanding of the double helix of DNA, which informs all aspects of our corporeal being.

The kabbalistic interpretation of Adam's wrongdoing assumes another dimension: "Adam's sin actually consisted of his severing the Tree of Knowledge from the Tree of Life. Had he united the Tree of Knowledge with the Tree of Life, good and evil would never have existed. It was only after Adam separated the two trees that the Tree of Knowledge manifested good and evil. Only then did the *klipot* gain access to the Tree of Knowledge and draw its energy from there."[210]

The Tree of Knowledge possessed the entire cosmos of *Kedusha* and *Tumah* in a potential state. The consciousness of *Tumah* contained sparks of the all-embracing unified whole that Adam was to heal. Had Adam restricted and followed the command of the Master of the Universe, the *tikkun* process would have been completed. The sparks would have been freed from the clutches of the Dark Lord and the kingdom of evil never would have become manifest. But because of Adam's transgression, *pesh'a* (crime) was established within *Malchut*. Furthermore, Adam's sin drew the serpent close to Eve on the terrestrial level because what happens Above also happens Below. Thus, the serpent came upon Eve and injected into her the poisonous venom of the Desire to Receive for Oneself Alone.[211]

The *zuhamah* (filth) by which the serpent defiled *Malchut* brought corruption to all of Creation. Purification of the cosmos will not occur until the Final Correction. The prophet Isaiah referred to the rehabilitation and restoration of the universe to perfection when he declared, "He will swallow up death in victory; and the Lord will wipe away tears from off all faces."[212] In Kabbalah, "eyes" refer to *Chochmah* and *Binah*, and "tears" symbolize the lack of *Mochin*, the Light of Redemption. This absence was a result of the *zuhamah* that became commingled with the Lower Cosmic Consciousness after Adam's sin of eating the fruit from the Tree of Knowledge. The prophet Isaiah, therefore, states that tears shall not disappear from *Malchut* until the Dark Force is abolished for all eternity. When *pesh'a* (crime) is eliminated, the Dark Lord will cease to exist.

Between cosmic *Atzilut* and cosmic *Briah* there exists a *parsah*, a cosmic boundary, similar to the Van Allen belt in our solar system. The Lord told Adam not to draw any Light below this cosmic boundary because Satan could seize any Light that entered the three lower cosmic universes. Had Adam obeyed, the Worlds of *Briah* (Creation), *Yetzirah* (Formation), and *Asiyah* (Action) would have been raised into *Atzilut* (Emanation), thereby restoring our mundane world to eternal perfection. However, because Adam did not raise the sparks and the three lower universes into *Atzilut*, none of Creation became perfected. Instead, Adam went down into the three lower Worlds, which were now united with evil, and he drew the *shefa* (abundance) below the barrier of *Atzilut* with him. These matters were encapsulated in the words with which the Creator answered *Pei*: "You are beautiful, but deeply hidden within you is the impression of *pesh'a* (crime)."

In other words, even though *Pei* embodies both *Pedut* (Redemption) and the Light of Wisdom, it is the *Mochin* issuing from *Ima* that will eventually be bestowed upon *Pei*, bringing redemption to the world. Thus, redemption is precipitated by the *Mochin*. Until the level of Messiah-consciousness is reached, all redemption that comes to pass in the world will lack completion. This is evidenced by the people of Israel being repeatedly driven out of their land and by the destruction of the two Holy Temples. This is all due to the fact that *Pei* contains *pesh'a*. The redemption that *Ima* can provide now to *Pei* is still unable to entirely redeem the *pesh'a* (crime) of Adam's sin. Consequently, *Pei* will always be at risk of being seized by the *klipot*.

When the Lord answered *Pei*, He also likened the letter's appearance to a serpent that bites a person and immediately hides its head within its body. This trait makes it impossible to kill the serpent because it can only be killed by being hit on the head. The *Zohar* tells how after making Adam sin, the serpent curled up his head within his body and "stretched out his hands."[213] Even though *Malchut* obtains the *Mochin* for redemption through *Ima* (*Binah*), the bent head of the serpent represents the Light of the Body rather than the Light of the Head.

The Light of the Body refers to the *Sefirot* of *Chesed*, *Gevurah*, and *Tiferet*, which are revealed by the *Mochin* and imparted to *Malchut*. The serpent thereby gains access to *Pei*, causing *Pei* to be without the *Mochin*, which is necessary for the creation of the world. This being so, the letter *Pei* is unfit to bring the world to its Final Correction and ultimate perfection.

From all of the above, we know two things about *Pei*. First, we know that *Pei*, which symbolizes *Malchut*, is contaminated with the venom of the serpent. Second, we know that the *Mochin* received from *Binah* does not supply *Pei* with enough power to overcome the enemy. This vulnerable condition of *Pei* shows that she receives the Light stemming from the Body and not from the Head of the *Sefirot*. Hence, the Lord told *Pei* she was not a fitting channel for Creation.

With head bowed, *Pei* departed from the presence of the Master.

Electricity;
Opposing Forces;
Rejoicing;
Distress;
Abundance;
Languages;
Cosmic Influences;
Logic;
Discrimination;
Misfortune;
A Holy War

CHAPTER 11

THE LETTER
AYIN

PEOPLE SEE THINGS AS
THEY ARE AND ASK WHY.
I DREAM OF THINGS
THAT NEVER WERE AND
SAY WHY NOT.

—JOHN F. KENNEDY

No mention is made in the *Zohar* of the letter *Ayin* approaching the Lord with an appeal to be the channel and seed for Creation. Cosmic *Netzach* (Victory), the internal life-force of *Ayin*-consciousness, reveals itself as the unseen companion of cosmic *Hod*. Much of the character of *Hod* is determined by its deep connection with *Netzach*. Just as the positive and negative poles of a light bulb jointly provide the conditions for the manifestation of electricity, *Netzach* and *Hod* do the same within their abode in the Lower Triad of the Star of David. *Netzach* and *Hod* are considered as two sections of one part of the Body of the *Sefirot*. They are, in effect, two opposing forces that together provide a unified expression of the all-embracing whole.

Ayin means "eye" in Hebrew. *Ayin* is the only letter within the *Alef Bet* that represents a specific physical object. Words beginning with the letter *Ayin* point to the predominance, beneficial or otherwise, of consciousness over physical matter. Just as stone and brick are used to construct buildings, words are assembled by the ordering and grouping of letters, which are the channels by which intelligence and intelligent energy-forces become manifest. The letter *Ayin*'s location within a word designates a particular state of consciousness. *Oneg* עֹנֶג, meaning "rejoicing,"[214] begins with the letter *Ayin*, indicating the achievement of an altered level of consciousness. However, when the letters of this word appear in reverse order, the word or intelligent energy emerging from this altered arrangement is *negah* נֶגַע, meaning "plague" or "distress."[215]

Because of the intimate relationship between *Ayin* and its companion letter, *Pei*, we find similar reasons why neither letter was used as the channel for Creation. *Pei* was not used because of the *pesh'a* (crime) that penetrated the inner recesses of her cosmic intelligence. *Pei*'s intimate connection with *pesh'a* is because *Pei* is the first letter of the word *pesh'a*. However, when the letters of *pesh'a* פֶּשַׁע —*Pei*, *Shin*, and *Ayin*—are rearranged, a new word is created: *shefa* שֶׁפַע, meaning "abundance."

The Hebrew language and the *Alef Bet* never reach the limits encountered by other languages. As Heisenberg put it: "The problems of language here are really serious. We wish to speak in some way about the structure of the atoms, but we cannot speak about atoms in ordinary language."[216] If we have reached the limits of ordinary language, are there routes of thinking beyond language that we can use to understand non-conceptual reality? Yes, say the kabbalists.[217] The study of the world of atoms forced physicists to realize that our usual language is totally incapable of describing atomic and subatomic reality, but the study of cosmic reality revealed no such limits to the Hebrew language.

The *Alef Bet* is no ordinary language. The entire known and unknown universes were created by means of the letter-energies contained in the *Alef Bet*. Just as life is a cycle, with varied negative and positive cosmic influences appearing at different times, so, too, do the letters of the *Alef Bet*, through words and syllables, act out series of cyclical events.

This was precisely the reason for *Pei*'s rejection. When we discover the indeterminate nature of *Pei*, we are faced with a cosmic paradox. *Pei* was the channel making manifest the intelligent energy-force of *pedut* (freedom). However, the interpenetration and mutual relationship of the intelligent energies of the letters could cause *Pei* to be the catalyst for the creation of *pesh'a* (crime) in our universe. Thus, there is within *Pei* a blending or confusion between the concept of crime and the concept of freedom. If we cannot correctly choose between these two concepts, then the process of *tikkun* (spiritual correction) has no chance of success. Therefore, another intelligent energy—one not possessed of this kind of either-or nature—would have to be found.

Scientists have striven for the past several centuries to discover the basic all-inclusive laws of nature. Today, however, the quantum physicist has come to view the universe—at least on the subatomic level—as a phenomenon that is co-created by the human mind. If this is true, then all human observations and conceptual constructs must be dealing with a conceptual map of reality, rather than reality itself.

For those who cannot accept a many-faceted reality, these new scientific advances open up a Pandora's Box of potentially severe consequences. After all, if there is no such thing as a single all-inclusive reality, then it seems that any one definition of what is real is no better than any other. And if reality itself is not real, then what of concepts such as freedom, justice, and honesty? Are they not also open to interpretation? Indeed, are not all human concepts now subject to the whims of those who would use them to serve their own ends? Who is to say what is right or wrong, now that reality itself is beyond understanding?

This was a dilemma that the Master of the Universe wished to avoid. He rejected the plea of the *Pei* because of the *pesh'a* (crime) within her framework. He did this despite the fact that *Pei* symbolizes the negative polarity, cosmic *Hod* of *Zeir Anpin*, and despite the fact that the *Mochin*, the Light of Redemption, also resides within her consciousness. Cosmic *Pei*, with her negative energy-intelligence of the Left Column, closely resembled the Dark Force in some respects. Thus, had *Pei* been accepted as the channel for the Creation, the possibility of suffering and misfortune would have been greatly increased. True, cosmic *Pei* also contained the energy-intelligence of freedom, but this intelligence could have been seized by the Dark Lord, dooming Earth's inhabitants to perpetual warfare. The Master was left with little choice but to reject cosmic *Pei* as the channel for Creation.

Then cosmic *Ayin*, the positive polarity, came forward to press her suit to be the channel for Creation. "While it is true," stated *Ayin*, "that *Pei*, as a negative polarity, drew the *Mochin* (Light of Redemption) from *Binah*, it is the Right Column and not the Left Column that expresses the energy-intelligence of the *Mochin*. I contain *Netzach*-consciousness, which is positive in nature. Consequently, I can make manifest the Light of Redemption. Furthermore, as the first letter-energy of the word *anavah* (humility),[218] I represent the nature of *anavah*. Thus, if I am chosen, the Earth's inhabitants would be positively influenced by the characteristic of humility, and the negative qualities of hatred, evil, and envy could be greatly reduced.

"Moreover," *Ayin* continued, "being the embodiment of cosmic *Netzach* further strengthens my ability to withstand the onslaughts of Satan. There exists between us a natural gap that does not exist between *Pei* and the forces of Darkness. The energy-intelligence of cosmic *Netzach* is positive, whereas the primary energy-intelligence of Satan is negative."[219]

The Master of the Universe replied, "But you, *Ayin*, also characterize *avon* (sin), as does your companion, cosmic *Pei*. The letter *Ayin* is the first letter of the energy-intelligence for *avon*."[220]

If *Ayin* became the channel for Creation, the Dark Lord would be given the opportunity to seize upon the first sin of mankind as a means to perpetuate his reign over the Earth's inhabitants. The principle of humility might never recover from such an attack. The concept of humility would then be vested only within the meek, poor, and innocent, and there would be no room for the strong to express the character of a humble human being.

Cosmic *Ayin*, with her head bowed in humility, departed.

CHAPTER 12

THE LETTER
SAMECH

BY THREE THINGS THE
WORLD EXISTS;
BY THE BIBLE, BY SERVICE,
AND BY THE DEEDS OF
LOVING KINDNESS.

—SHIMON THE RIGHTEOUS, *PIRKEI AVOT* (ETHICS OF THE FATHERS)

The search for the cosmic bond continued. Despite cosmic *Ayin*'s rejection by the Master of the Universe, cosmic *Samech* felt qualified to become the channel through which the world would be established. Yet she was hard-pressed to identify some fundamental reason why she would be a more appropriate instrument for the grand unification than any of the other letter-energies. In search of facts she could cite before the Lord, *Samech* began to reflect upon the very infrastructure of His dominion, the mystery of life itself.

By delving deeply into the celestial and terrestrial levels of consciousness, cosmic *Samech* hoped to establish a stronger case than those who had gone before her. If *Samech* could understand how the universe would maintain the balance of all energy life-forms, she could then decide how her own unique energy-intelligence qualified her for the assignment.

Let us, with *Samech*, ponder the most profound mystery of them all: the origin of life. According to scientists, primitive life-forms began to emerge on Earth at least three billion years ago. Some progress has been made in recent years in uncovering certain basic principles that seem to control the appearance of life, but as yet, the riddle of life remains unanswered.[221] Charles Darwin is famous for postulating the concept of biological evolution, but while his findings shed light on how life developed on Earth, they did nothing to illuminate the origin of life. Even Francis Crick, who co-discovered DNA, professed ignorance concerning life's primordial origins. In Crick's own words, "It is impossible for us to decide whether the *origin* of life here was a very rare event or one almost certain to have occurred. It seems almost impossible to give any numerical value to the probability of what seems a rather unlikely sequence of events." Despite our best efforts, it appears that the more we learn about our origins, the less we know.

Is life the result of random chemical activity? Or is humankind the expression of Divine Will? According to the *Zohar*, human life, symbolized by Adam, represents the crowning achievement of the Lord's cosmic blueprint.[222] "So the Lord created man in His own image," states the Bible.[223] This verse asserts quite explicitly that life is the direct result of the Creator's concerted activity.

The *Zohar* provides further commentary on the origin of life: "Rav Chiya commenced to discourse on the verse: 'The flowers appear on the Earth, the time of song is come, and the voice of the turtle is heard in our land.'[224] He said, 'When the Lord created the world, He endowed the Earth with all the energy requisite for it. It did not put forth fruits until man appeared. When, however, man was created, all the products that were latent in the Earth appeared above the ground. Similarly, the celestial bodies did not impart energy-intelligence to the Earth until the emergence of man.'

"So it is written: 'All the plants of the Earth were not yet sprung up, for the Lord had not caused it to rain upon the Earth and there was not a man to till the ground.'[225] The harvest of the Earth was still hidden in its bosom and had not shown itself. The Heavens refrained from pouring rain upon the Earth because man had not yet been created. When, however, man appeared, the flowers sprang forth from the Earth, and all its latent powers were revealed."

That "the time of song had come" was indicated by the Earth now being ripe to offer praises to the Lord, which it could not do before man was created. "And the voice of the turtle is heard in our land" indicates, according to the *Zohar*, that the energy-intelligence was not present in the universe until man was created. When man emerged, so did life on Earth. When man sinned, the Earth was cursed, and all these good things left it. As the Bible states: "Cursed is the ground for your sake,"[226] and again: "When you till the ground, it shall not give its strength to you."[227]

There has been much speculation about the possibility that life may exist elsewhere in the universe. No hard evidence has been found to prove the existence of extraterrestrial life forms, but most scientists believe that there may be life on other planets, given the age and vastness of the universe. Hopes for solving this mystery rose with the advent of space exploration, but when man's landing on the moon offered no clues, many of those hopes were dashed. Moon rocks taught us more about the universe, but nothing about the possibilities of other life in the universe. The unmanned ships that landed on Mars taught us even more, but they, too, added nothing to our store of knowledge about the mystery of life.

According to the *Zohar*, Adam was the crowning achievement of the Lord's cosmic blueprint. Adam the man is the earthly symbol of Adam Above—*Adam Kadmon* (Primordial Adam). Adam symbolized the material form of all Creation. Yet for its awakening, the emergence of human life would have to await the Sixth Day.

All activity includes one of two forms of restriction: voluntary and involuntary. Involuntary restriction prevails at all levels of earthly and celestial existence, including many of the bodily functions of man. But there is one exception to the rule of involuntary restriction—the will of man. Only humankind's metaphysical components—our minds and souls—have the option of exercising voluntary restriction and thus revealing infinite energy-intelligence. The rest of Creation operates on a more or less automatic, or reactive, level. When we choose restriction, we reveal the origins and dynamic interplay of life itself; when we choose not to restrict, we remain in a world of darkness and illusion. The creation of Adam triggered a ceaseless flow of energy manifesting as intelligent life-forms, and the whole of Creation then came alive.

The Lightforce, accompanied and channeled by the energy-intelligence of the *Sefirot*, first became manifest on the Third Day of Creation. The biblical Third Day, according to kabbalistic interpretation, indicates the emergence of the Central Column that unites the two opposing energy-intelligences of the Lightforce, namely the Desire to Impart (Right Column) and the Desire to Receive (Left Column). The conductor between these opposing forces is the energy-intelligence of restriction. When restriction takes place, the Lightforce is kept from entering the Vessel of the energy-intelligence of Desire to Receive, thus causing an explosion that reveals the energy-intelligence of Desire to Impart.

Imagine a universal blackout in which all cosmic activity is slowed to a virtual standstill. This was the condition before the appearance of Adam. All systems were in a state of dormant suspension, ready and waiting for the moment when the switch would be turned to the "on" position. When the energy-intelligence of Adam (man) appeared, the infinite Light of Creation illuminated the Heavens in a colossal explosion, and the web of life began its cosmic dance.

The *Zohar* explains the present expansion of the physical universe as due to the thrust of Adam's initial explosion, brought on by the original restriction.[228] According to this model, humankind (which evolved from Adam), continues to trigger this activity of the cosmos. Cosmic expansion and contraction depend on whether or not we choose to restrict. When we restrict, we infuse positive energy-intelligence into the cosmos. If we choose instead to surrender to the Desire to Receive for Oneself Alone, we generate cosmological upheaval.[229]

The *Zohar* declares that an understanding of Adam will provide us with a full comprehension of the cosmos. Noted kabbalist, physician, and astronomer Shabbatai Donolo states in his commentary on the *Book of Formation* that man is a reflection of the entire cosmic maze of physical properties and metaphysical interactions. It was to this issue that *Samech* now applied herself. Could she be a just and proper channel for an orderly universe? Was her particular metaphysical framework suited to assisting man when he became consumed by his own selfish desires?

Before continuing to describe the plea of *Samech*, we must examine the interplay of energies during and immediately after Creation. By so doing, we will better understand the essence of cosmic *Samech*. The energy-intelligence of the *Samech* in the *Alef Bet* lies within the *Sefira Tiferet*, which is represented by the Third Day of Creation.[230] The Third Day, up to now, has been obscured by the cosmic code in which it is encrypted. The biblical version of the Third Day is vague about exactly what took place on that day. However, the *Zohar* says that "the Third Day of Creation is the code name for the *Sefirotic* energy-intelligence of cosmic *Tiferet*. The *Sefira* of *Tiferet* encapsulates the dominion of the Central Column over the other two energy-intelligences." It is for this reason, kabbalistically speaking, that the number three denotes the necessity for three elemental forces necessary in our universe.[231]

The author of the *Zohar*, Rav Shimon bar Yochai, condemned the popular interpretations of the Bible. According to Rav Shimon, the Bible is not a detailed account of religious doctrine; rather, it is nothing more than a complete code of our cosmology. The Bible speaks in code; the Kabbalah does the deciphering.

What seems to emerge from the biblical account of Creation, and its supportive interpretation by the *Zohar*, is that the Earth and its entire physical framework came into existence on the Third Day of Creation, which is represented by the *Sefira* of *Tiferet*.[232] The physical universe came into being when the third force, the mediating principle of restriction known to Kabbalah as the Central Column, came into existence. Man and the animal kingdoms followed at a later date.

According to Kabbalah, the universe has been programmed to evolve in a series of Ten Emanations toward its Final Correction. A uniform, well-arranged order, guided by an innate metaphysical pattern, encompasses all phenomena, from the simplest to the most complex. Just as DNA provides all of the ingredients necessary for biological duplication and the transmission of hereditary properties, so does the metaphysical world function according to ten inherent interactions. The *Zohar* suggests there is a system that functions much like the biological DNA system. This metaphysical DNA concerns itself with natural phenomena, determined and influenced by the over-all design in nature. Like Crick's genetic molecule, Earth and her inhabitants contain unique metaphysical patterns that are the metaphysical equivalent of DNA. The *Zohar* points out that the behavior and orientation of any individual entity is not only genetically conditioned and predetermined, but it is also the result of the position of its own metaphysical DNA within the cosmic whole.

We already know that the full development of man depends on his establishing a proper relationship with his physical environment. The same is true at the metaphysical level. Rav Shimon said, "As long as the community is complete and joyful, blessings abide to all. However, when the world is not connected with the all-embracing unity, then blessings are withheld by Him and from all others."[233] The *Zohar* says we can shape the relationship between man and the environment for good or for ill, according to the criteria we choose.[234] This shaping can take place at both the mundane and the metaphysical levels. Man has the ability to modify his metaphysical DNA, and all successive generations will inherit the modification.

The code and structure of our metaphysical DNA lie within the framework of the *Sefirot*. In our discussion of the seven days of Creation, bear in mind that we are referring to activity on a pure energy-intelligence level—the level of metaphysical DNA. Our frame of reference for the actual establishment of the whole of our universe necessarily excludes any reference to the physically expressed universe, which emerged later, as is indicated by the two accounts of Creation in Genesis.

Earth, too, has its own complete individual cosmic code of seven encapsulated intelligent forces—the Lower Seven *Sefirot*. These primal forces are responsible for the solar system and for Earth's own cosmic division. The seven channels of energy-intelligence that emanated from the seven *Sefirot* during the time of biblical Creation were, and still are, directly responsible for Earth's various geographic manifestations.[235]

Because the creation of Earth preceded the creation of the celestial bodies, Earth is considered to be the cosmic seed of our solar system. The planets arose from a replication of Earth's complement of information. Earth is cosmic *Tiferet*'s channel of energy-intelligence. Each *Sefira*, or channel of energy-intelligence, consists of its own seven metaphysical DNA subdivisions, just as the seed of a tree contains the DNA codes for all the sub-components (roots, leaves, and so on) of the entire tree.

When a tree produces an identical copy of itself, it relies on multiple forms of expression. So it was with Earth during the initial period of Creation. Seven encapsulated energy-intelligences burst forth from cosmic Earth, seeding the solar system with seven complex energy-intelligences, manifested by the energy-intelligence of cosmic *Netzach*'s framework. This brought into existence the seven celestial bodies: Saturn, Jupiter, Mars, the sun, Venus, Mercury, and the moon.

In the biblical description of the Fourth Day of Creation, the term "day" should not be taken literally. "Day" is code for *Sefirotic* energy-intelligence. The Third Day, deciphered, is the cosmic intelligence of *Tiferet*, and the Fourth Day is the cosmic energy-intelligence of *Netzach*. What emerges from this interpretation of the Bible is a dynamic process of *Sefirotic* evolution. This sheds a new and more comprehensive light on the question of what came first and what came later.

While we are on the subject of evolution, I would like to address Darwin's theory, a controversial subject that shook the very foundations of Judeo-Christian doctrine and became the battleground for history's greatest ideological dichotomy, pitting religion against science. The evolutionary process suggested by the *Zohar* raises some serious questions regarding the origin and evolution of human life. The biblical account of Creation indicates that plants, animals, and man appeared on earth in a sequential manner.[236] This might be taken as confirmation of Darwin's theory that living things change from one form to another as a result of random events, and not from deliberate intervention on the part of a Divine power. A *Zoharic* examination of the master plan of Creation yields a more profound appreciation of the biblical accounts of Creation.[237] Close inspection of Genesis reveals the act of Creation to be repetitive. The creative activity of the second chapter seems to be identical to that in the first chapter, but according to the *Zohar*, the account of Creation in the first chapter represents Creation when it was still in a potential state, while the account of Creation in the second chapter refers to the physical Creation.

The natural laws of the metaphysical realm ensured that the first Vessels to emerge after the *Tzimtzum*, or First Restriction, were those with a higher degree of purity, and consequently with a lesser degree of the Desire to Receive. Vessels are the exact opposite of Lights. The former are the expression of the Desire to Receive, the negative aspect, while Lights are the energy-intelligence of the Lightforce, the Desire to Impart, the positive aspect.[238] Because the first Vessels had a lesser Desire to Receive, the first Lights to emerge were, correspondingly, those with a lesser degree of the Desire to Impart.

The paradox in this process is that the supreme level of the Lighforce, *Yechidah*, cannot be expressed until the lowest, most intense energy-intelligence of the Desire to Receive is made manifest.[239] This paradox illustrates another important characteristic—perhaps even the essence—of the kabbalistic worldview: the knowledge that all seemingly contradictory and irreconcilable concepts and entities are aspects of a single basic unity.

The World of Action represents the lowest and final Vessel of energy-intelligence and the highest degree of Desire to Receive. It is only by virtue of the expression of this negative aspect of desire that the supreme level of the Lightforce becomes manifest. Understanding this relationship provides us with a deeper comprehension of the paradoxes facing society today. On the one hand, we observe a multitude of physically oriented, pleasure-seeking, non-spiritual people; on the other hand, we see a reawakening of spirituality in diverse places and cultures all over the world. This coexistence of opposite intentions and desires lies at the heart of the kabbalistic view of the Age of the Messiah.

Now let us return to the mystery of biblical Creation. The *Zohar* tells us that the first account of Creation in Genesis illustrates the evolution of Vessel energy-intelligence, that is, the emanation of the *Sefirotic* process, whereas the second account describes the birth of the physical world. The evolutionary cosmic intelligences of Day One through to the Sixth Day are the *Sefirot* of *Chesed*, *Gevurah*, *Tiferet*, *Netzach*, *Hod*, and *Yesod*, and each of these contains a different degree of the Desire to Receive, each *Sefira* containing a more intense manifestation of the Desire to Receive than the one preceding it.

The first account of Creation, therefore, describes the metaphysical emergence of man as the ultimate Vessel for the unveiling of the Lightforce. The Desire to Receive attained its most intense manifestation within man. Man's Desire to Receive, the greatest such desire in the universe, provided the Lightforce with a Vessel conducive to revealing the greatest abundance of Light. The greater the capacity of the Vessel, the greater the intensity of the Lightforce that fills it. The second account of Creation in Genesis describes the manifestation of the physical universe and man as a corporeal entity, the crowning articulation of the Lord's activity, with the ability to dominate extraterrestrial activity from the terrestrial level.

The term "animal" is the code for cosmic *Hod* and its level of Desire to Receive, which was much smaller than that of cosmic *Yesod*, man. Corporeality is not the subject dealt with in the first description of Creation because animal and man did not yet exist within a corporeal framework. The energy-intelligence of the sixth *Sefira*, *Yesod*, was still in a pure state of potential energy.

"The Lord created man in His own image," states the Bible. In fact, all physical manifestations contain a degree of the Creator's life-force, but the Lightforce found its full manifestation only in man. Therefore, in the second chapter of Genesis, physical man appeared before the animals; animals being a lower level of consciousness. We now have a deeper understanding of the verse that states: "And Adam called all the animals, birds and all creatures of the field by their names." The word "call" is code for the control of energy (see the chapter dealing with the letter *Kuf* for a fuller explanation of the word and concept). Adam (man), the crowning achievement and manifestation of the Lightforce, was given the ability to control energy, as stated by the *Zohar*.

This is why the stability of the universe would depend upon man's conduct. Man embodies the duality manifest in the universe. Both polarities, negative and positive, would be expressed by man's actions. Violence, war, and suffering would be of his own doing, just as peace on Earth and goodwill toward his fellow man would depend entirely on the degree of restriction with which man confronted the Lightforce.

After some consideration, cosmic *Samech* now understood the requirements for an orderly universe, and she concluded that she was by far the fittest instrument for the Creation of the universe. So she approached the Master of the Universe and said, "Lord of the Universe, may it please You to create through me the world, inasmuch as I represent the function of *smicha* (supporting) the fallen ones. You, Lord, portray this characteristic as stated in the verse: 'The Lord upholds all that fall.'"[240]

To avoid mishaps in man's attempt to achieve his *tikkun*, the Earth and its environment must be cosmically stable—and stability was cosmic *Samech*'s objective. She knew perfectly well that harmony would be maintained in the world if she were chosen to be the channel for Creation. The letter *Samech* believed that she alone of all the letter-energies could meet the requirements expressed in the Scripture that reads: "Do you know the ordinances of the Heavens? Can you establish their rule on Earth?"[241]

The kabbalists call the energy necessary to maintain a balanced cosmos the *Or DeChochmah*, the Light of Wisdom. The structure of the *Or DeChochmah* is the kabbalistic equivalent of the structure of the atom, the intrinsic schema governing all energy, the initiating Lightforce that is the seed of atomic energy. The discovery that atoms are composed of three kinds of elementary particles—protons, electrons, and neutrons—has brought modern physics closer to the Kabbalah. This basic triad, in different combinations, forms the basis for everything that exists.

Everything, material and beyond, is composed of atoms. Your hand moves freely through the endless sea of atoms around you, yet it stops on contact with a table. Why should this be so? The answer lies in the nature of the electron, which has the negative energy-intelligence of Desire to Receive for Oneself Alone. The surface of an atom is composed of electrons that exhibit negative energy-intelligence, so all other structures, whenever they appear, are repelled by that negative energy. Hence, our hands do not pass through rock, our bodies cannot walk through walls, and we do not fall through our beds at night.

Deep inside the atom is the nucleus, which is a hundred thousand times smaller than the atom. Recently, physicists have come to believe that protons and neutrons are composites of other subatomic particles, which in turn appear to be composites of still smaller particles. The endless chase for smaller composites is on. Infinity has intruded into the once-manageable haven of physics. All that remains is uncertainty.

Will we ever complete our understanding of nature? Is there truly an infinite regression into more and more basic particles? In the view of the kabbalist, the answer is yes because energy is intelligence, and intelligence is found within the realm of the infinite world of metaphysics. Where do the naturally occurring elements come from? How did the energy-intelligence of such power evolve? The *Zohar* sheds light for us with the following passages:

> *Rav Aba began his reflections of this portion with the verse:*[242] *'Trust the Lord forever (ad), for in Yah (Yud-Hei), Yud, Hei, Vav and Hei—the Tetragrammaton—is the fashioning of the universe.' All mankind should cleave to the Lord and put their trust in him. This, so that their strength will be drawn from the sphere called Ad (Tiferet), which sustains the universe and binds it into an indissoluble whole. The Central Column, Tiferet, is the binding energy-intelligence, the mechanism that joins the Right Column (proton) with the Left Column (electron). This Ad is the desire of the everlasting hills[243] of Binah and Malchut. Binah is the source from whence all blessings originate, and her desire to*

crown the Lower Worlds with blessings requires Ad, or Tiferet, in her transmission. Malchut, in turn, longs to receive these same blessings and energies.

Therefore it says, 'Trust the Lord unto Ad,' [that is,] contemplate the Worlds of Emanation only as far as the level of Ad. For beyond that level is a hidden region, so transcendent that it passes all understanding—the very source whence the worlds were designed and came into being. Up to this point, it is permissible to contemplate the threefold unity, for it is wholly recondite.

"This is Yah, from whence all worlds were fashioned," said Rabbi Judah. "We have direct scriptural evidence for this, for it is written:[244] 'Ask of the days that are past—since the day that the Lord created man upon Earth—and ask from the one side of the Heaven unto the other.'"[245]

The *Zohar* reveals the Master's Superstructure: the *Yud-Hei* of the Tetragrammaton (*Binah*), the innermost aspect of the Lightforce. The *Yud-Hei* of the Tetragrammaton remains forever concealed and beyond human comprehension. Man may investigate up to this point, but no further. However, when the Lightforce enters the framework of *Tiferet*, the Light is transformed; the initial three particles of energy, or aspects of the Lightforce—the Upper Three *Sefirot* of *Keter*, *Chochmah*, and *Binah*—assume a change in names to indicate this lower level of energy-intelligence, and their Light becomes a unified whole.

The energy-intelligence of cosmic *Samech* is the *Sefira* of *Tiferet*—the bonding ability that brings opposites together within the reality of the Lightforce. In a sense, the *Sefirot* of *Chesed*, *Gevurah*, and *Tiferet* might be considered elementary particles. They are the structures that make up the unknowable *Yud-Hei* and therefore are the ultimate in energy-intelligence. *Tiferet* is the Lord's way to support and balance all subsequent Emanations, or *Sefirot*, at the most elemental level.

Consequently, the letter *Samech* felt that her energy-intelligence was necessary for stability in the world of *Asiyah*, the corporeal manifestation of the Lord's Kingdom. In fact, the Lord had already decided that the Earth would be created on the Third Day for this very reason: The Third Day was an Emanation of the Central Column, of which cosmic *Samech* was the primary source.

If man were to corrupt his deeds, he would cause the *Mochin* (the Light of Redemption) to be expelled from *Zeir Anpin*, the overall framework of *Tiferet* and *Malchut*. This in turn would reduce the Lightforce within *Zeir Anpin* and *Malchut*. To prevent this outcome and avoid catastrophic violence in the cosmos, the Lord provided the channel of energy-intelligence known as *Samech*. Cosmic *Samech* was the Master of the Universe's connection to His Superstructure, the *Yud-Hei*, also referred to as *Da'at*. As mentioned previously, cosmic *Samech* was called the Light of Mercy—a step down from the Light of Wisdom, but nevertheless an energy-intelligence distinguished by its astonishing primal power. This is why the Light of Mercy is called *Avira Dachyah* (Pure Atmosphere). Cosmic *Samech* was on a par with the Light of the Lord's Superstructure.

Man's wrongdoing could lead to a state of universal imbalance. However, no *klipa* (evil energy-intelligence) can compete with the conscious energy level of *Zeir Anpin*, the whole of *Tiferet* and *Nukvah* (*Malchut*). If imbalance occurred, cosmic *Samech* would step in to protect, support, and maintain balance within the cosmos.

The word *samech* signifies "support." Cosmic *Samech* protects the conscious levels of *Zeir Anpin* and *Malchut* and helps them maintain their positions within the Master of the Universe's realm of cosmic energy-intelligences, thereby preventing the Dark Lord from establishing a base within the Master's domain during the period of immaturity. Earth is a composite of the Third Day, the power of restriction, and the neutron-consciousness of the Lord's Superstructure, the *Yud-Hei*, which cause Earth to be closely aligned with the energy level of the *Samech*. This is why cosmic *Samech* would have no difficulty coming to Earth's rescue during its periods of immaturity; the similarity between them provides a linkage between their levels of consciousness.

Thus, *Samech* considered herself to be far better suited for the Creation of the universe than any of the preceding letters. The energy-intelligence of *Samech* achieved an affinity with the essence of the Lord's Superstructure, causing the immense power of *Samech* to remain untouchable by the power of the Dark Lord. When the need arose, cosmic *Samech* could descend into the province of *Zeir Anpin* and *Malchut*, the level of consciousness to which the security shield of the Lord's Superstructure did not extend.

Although that security shield could not protect *Samech* during her descent, she had no fear of the Dark Lord. The preceding letters were susceptible to capture by Satan once they left the protection of the Master, but not cosmic *Samech*. Her energy-intelligence originated in the Upper Realm, the Superstructure of the Lord. Satan was no match for cosmic *Samech* and he knew it. The power of *Samech* could disintegrate the entire realm of the Dark Lord. Cosmic *Samech* could defeat Satan even when *Zeir Anpin* and *Malchut* were in a weak, immature state of Cosmic Consciousness due to humankind's corruption of the universe. In other words, *Samech* felt she could protect man even when his own deeds rendered him vulnerable to attack. Earth and the entire universe had a better chance with cosmic *Samech* as the instrument for Creation.

After *Samech* stated her case, the Lord replied, "Your own reasoning dictates that you remain at your cosmic conscious level and not leave your position unmanned."[246] What did the Lord mean by this? He was referring to the fact that *Zeir Anpin* and *Malchut* were inclined to fall from their cosmic level of consciousness due to man's free will. But if the universe became pervaded with *Samech*'s energy-intelligence, then permanent immaturity would become a real possibility because man would find it unnecessary to bring himself to the level of *Tiferet*-consciousness. There would be no impetus for humanity to manifest the restriction necessary to remove Bread of Shame. Only with Earth's environment and inhabitants unaware of *Samech*'s cosmic presence could free will prevail.

The energy-intelligence of *Mayin Nukvin* (Returning Energy-Intelligence) maintains and supports the entire cosmos.[247] If we lived on a planet where there was little change, boredom would set in. There would be no motivation for mankind to improve. On the other hand, if our universe were an unpredictable one where things changed haphazardly, we would have no way of knowing what to do next. We live in a fluctuating universe because humankind is always in a state of modification and variation. We create this condition by our own exercise of free will. Yet there is order and a kind of predictability in the cosmos; planets still move in predictable orbits, displaying a precision more refined than even clockwork can boast.

Violence has its role in the history of nature. Quite apart from the myriad storms, upheavals, and disasters in the history of the Earth, radio telescopes frequently receive signals from dying stars and colliding galaxies. Then again, the outlook appears more serene when we consider that for almost every known peril, a protective intelligence provides a remedy. For example, our ionosphere holds back destructive ultraviolet rays and other harmful radiation. A magnetic screen keeps cosmic rays under control. Earth itself is situated to the correct distance from the sun to assure it of the right measure of heat so that our water supply neither evaporates nor freezes in its entirety, instead remaining in the optimal state to sustain life.

In such surroundings, living forms enjoy prosperity. Sometimes, however, man plays dangerous games. He forgets his purpose and is a victim of amnesia regarding his own recent past. Cracking open the atom is but one such game that assuredly threatens man with his own destruction. The *Zohar* repeatedly strikes a discordant note on the interference of elementary forces throughout history. "These are the generations of the Heavens and the Earth," says the *Zohar*.[248] The expression "these are" denotes that those generations mentioned before are henceforth of no account. The Kabbalah refers to the products of *tohu* (emptiness), hinted at in the first chapter of Genesis: "And the Earth was *tohu* and *vohu*[249]: These it is of which we have learnt that the Lord created worlds and destroyed them."[250]

The lunar surface offers eloquent testimony to a previous age of destruction. Relative to the infinite reaches of space, the Earth is very near to the moon. Yet the moon is severely cratered by catastrophic collisions. Why has Earth been spared these destructive events? Could it be that *Samech* provides a cosmic shield? The *Zohar* says yes.

"If you, cosmic *Samech*, were to leave your position," said the Master of the Universe, "your absence would cause Heaven and Earth to remain in a balance forever precarious. They would constantly feel your presence; they would always need your support."

"Consequently," the Lord concluded, "because mankind would always depend on your support, and because *Zeir Anpin* and *Malchut* would lack the strength to stand up on their own, your energy-intelligence renders you unfit to be the channel for the Creation of the world."

With this, cosmic *Samech* left the presence of the Lord.

Lower Worlds;
Terrestrial Realm;
Inhumanity;
Pain;
Suffering;
Eruptions of Energy;
Collisions;
Security Shields;
Righteous;
Battle Stations;
Free Will;
World of Action;
Awesome Power

CHAPTER 13

THE LETTER
NUN

WE SAY THAT
INSEPARABLE QUANTUM
INTERCONNECTEDNESS
OF THE WHOLE UNIVERSE
IS THE FUNDAMENTAL
REALITY.

—DAVID BOHM AND BASIL HILEY

H aving observed and pondered the rejection of *Samech*, the letter-energy *Nun* entered the forum of energy-intelligences to plead her merits. *Nun* considered herself well suited for use in the Creation of the world, for she enjoyed a status as exalted as cosmic *Samech*'s.

The *Sefirot* triumvirate—the Upper Three *Sefirot*: *Keter*, *Chochmah*, and *Binah*—the Superstructure of the Lord, is in a state of constant productivity. Like a terrestrial power station, these first three *Sefirot* produce energy, regardless of whether their energy is used or not. They are the source of all cosmic energies.[251] But as these energy-intelligences of the Superstructure descend and enter *Zeir Anpin*, they undergo a transformation that reduces their energy-intelligence to an inferior state of consciousness. No longer are they referred to by their original *Sefirotic* designations. In place of *Keter*, *Chochmah*, and *Binah*, they are given new names: *Chesed*, *Gevurah*, and *Tiferet*, respectively.[252]

Cosmic *Samech* provided the energy-intelligence of the Central Column—the mediating restrictive principle referred to as *Tiferet*—to the Lower Worlds of *Zeir Anpin* and *Malchut*. The coupling of these two Lower Worlds was the secret of harmony and balance in the terrestrial realm. When humanity became endowed with the energy-intelligence of the Central Column, which is the Desire to Receive for the Sake of Imparting, the cosmic bond between *Zeir Anpin* and *Malchut* prevented chaos at all levels of the cosmic whole.

How did man come to merit the gift of the cosmic consciousness of *Tiferet*? By using his own latent energy-intelligence of *Tiferet*, which is inherent within him from birth until death. This energy is capable of restricting the powerful force of Desire to Receive for Oneself Alone. "For the inclination of man's heart is evil from his youth," states the Scripture.[253] Human activity activates or disengages the binding force, *Tiferet*. Balance and harmony at all cosmic levels depend completely upon cosmic *Tiferet*. Violence anywhere, celestial or terrestrial, emerges from man's inhumanity to his fellow man. Man's negative actions are catalysts for anger, war, and suffering.

The universe was born in a burst of violence known as the Big Bang. Today the universe is still rife with violent activity—huge eruptions of energy from agitated galaxies, horrifying upheavals and collisions—but this is not the Creator's doing. Rather, it is a reflection of man's activity on Earth. It should be noted that there was an element of competition among the letter-energies, a struggle that continues to this day. However, the letters are more or less equally matched, and unlike the human struggle, none of the letters ever attempted to overwhelm another. It is the actions of man, not those of the letter-energies, that sow the seeds of destruction. To counteract these forces, the Lord searched for an energy-intelligence that would spread universal harmony and tranquility. The Lord's search led Him to cosmic *Nun*.

Cosmic *Samech* could not leave her station in the Master of the Universe's Superstructure to be a permanent fixture in the created universe, for then man would never test the responsibility of restriction. The energy-intelligence of cosmic *Nun*, however, as a channel for the energy of *Gevurah* of *Zeir Anpin*, seemed ideal for the Creation of the universe. Her rank, similar to *Samech*, granted her the benefit of the security shield of *Zeir Anpin* of *Binah*. Hence, she had not the slightest fear of attack by the Dark Lord. Furthermore, she was not burdened with the responsibility of serving the Lower Worlds of *Zeir Anpin* and *Malchut* during the time of their immaturity. That task was left to cosmic *Samech*.

States the *Zohar*: "The *Nun* is the first letter and cradle of *Norah Tehilot* (Fearful in Praises) and *Navah Tehilah* (Praise for the Righteous)." Translating the unusual coding of *Nun*'s particular energy-intelligence into meaningful communication has been no trifling assignment. Nevertheless, Rav Ashlag deciphered the abstruse section of the *Zohar* dealing with this subject and provided an interesting interpretation of cosmic *Nun*'s energy-intelligence and of her subsequent plea to be considered the proper instrument for the Creation of the universe. According to Rav Ashlag's interpretation, when the intrinsic nature of the cosmic whole becomes threatened with disintegration or annihilation, cosmic *Samech* steps in. Cosmic *Samech*'s entry into the terrestrial realm drives the *klipot* back to their native environment. The Superstructure of the Lord, expressed in its own frame of reference, is thus transformed to another framework, and a new element arises: time—the Light of Compassion—is added to the three spatial coordinates of *Keter*, *Chochmah*, and *Binah*.

This new paradigm, known as time, provided the cosmic glue that bound the universe into a dynamic, unified reality. Satan lacks the Light of Compassion, the fourth dimension (time). The cosmic code word for this fourth dimension is *Tehilah* (Praise). If not for man's negative activity and his lack of compassion, Satan would have no place within the universal framework. Inhumanity provided the Dark Lord with a conduit into the basic unity of the universe, a necessity for the continuance of free will.

Binah, the Left Column energy-intelligence of the Superstructure, is the epitome of the Light of Compassion. In expressing *Binah*'s cosmic power, the *Zohar* refers to her as comprising *Nun Sha'arei Binah*, the Fifty Gates of Intelligence. Here we should note that the numerical value of *Nun* is fifty. Also, *Gevurah* of *Zeir Anpin* uses cosmic *Nun* in making her energy-intelligence manifest. *Nun* is an integral aspect of the triad that consists of *Mem*, *Nun*, and *Samech*. The *Sefira* of *Gevurah* is a dynamic force in *Nun*'s arsenal. *Nun*'s cosmic connection to *Tehilah* (Praise) is referred to as *Norah Tehilot*. Therefore, cosmic *Nun*, which inherently participates in the Light of Compassion and is referred to by the code name *Norah Tehilot*, would permanently serve to balance the universe and ensure universal harmony among the players on the cosmic stage.

Nun's plea to the Lord began as follows: "Master of the Universe, as I represent the Left Column of *Zeir Anpin*, the energy-intelligence of attraction, I am primarily responsible for the expression of *Tehilah* (Praise) by cosmic *Samech*. I therefore ask that you create the world through me. Due to the strength of the Light of Compassion that *Gevurah* received from *Binah*, the cosmic whole of *Zeir Anpin* of *Binah* achieved the name *Norah Tehilot*[254]—and I am the first letter-energy of the word Norah."

Nun continued, "I am also referred to by the coded verse: '*Navah* (Comely) praise by the *tzadikim* (righteous).'[255] Thus, cosmic *Tzadi*—*Yesod* of the Lower Triad of *Zeir Anpin*—gains her maturity by means of my energy-intelligence."

Nun represents the female aspect of the letter *Tzadi*, which means "righteous." Therefore, *Nun* is given the name of "Praise by the Righteous." As the energy-intelligence for the *Sefira* of *Yesod*, *Tzadi* provides spiritual nourishment to *Malchut*, the World of Action. Cosmic *Yud*, the vehicle for the Lord's Lightforce requires *Nun* for its manifestation. *Nun*, therefore, would be called upon to participate and function as *Yud*'s channel of communication with the World of Action. Her relationship with the terrestrial realm was thus established even before the world's inception. Consequently, *Nun* stated that she would be the proper instrument for the Lord to use to create the universe, for she would ensure that the cosmos would maintain stability and harmony.[256]

The Lord replied, "*Nun*, return to your place. It was for your sake that cosmic *Samech* was rejected. For you symbolize *noflim* (falling) because the word *noflim* begins with the letter *Nun*."

The Lord intended his reply to take cosmic *Nun* one step closer to grasping the total plan of Creation and to help her come to terms with her unique position within the cosmos.

The quality that predominated in *Nun*'s consciousness was the creation of harmony in the universe, but she failed to grasp the drama of the Final Redemption. Each letter would be a participant in this cosmic drama, not an outside observer. Indeed, the interrelationship of the letters with each other and with the cosmos would give meaning to the whole. Thus *Nun* was misguided in viewing Creation simply in terms of her status as guarantor of cosmic harmony. She neglected to take the other side of her twofold nature into account.

You might say *Nun* was something of a double-edged sword. While she occupied a unique and powerful position within cosmic *Gevurah*, she also played the role of cosmic *Yesod*, the *Tzadi*. Satan could not survive an onslaught by cosmic *Nun* in her aspect of cosmic *Gevurah*, which was why she pleaded to be used as the channel for Creation. But if *Nun* were to be called upon to assist in providing her energy-intelligence for cosmic *Tzadi* in the performance of *Tzadi*'s function on the stage of Creation, she would become exposed to attack by the Dark Lord.[257]

Cosmic *Nun*'s position within the *Tzadi* might require assistance were she to fall (*noflin*) prey to Satan. Once the Dark Lord penetrated a station within the security shield, there would be no telling what might occur. When *Samech* was told to return to her battle station in the event that there were *noflim* (fallen) victims of the Dark Lord, the Master of the Universe was already hinting at *Nun*'s potentially fatal flaw. *Nun* did not fully grasp the implications of *Samech*'s rejection. She saw only the positive side of her own energy-intelligence as the channel of *Gevurah*'s awesome power, which could provide Earth's inhabitants with the strength to withstand any attack by Satan and ultimately could usher in the Final Redemption of the world.

As the Lord said, it was no accident the first letter of *noflim* is the letter *Nun*. Words, like letters, express essential modes of existence and levels of energy-intelligence. Thus, if the letter *Nun* had been allowed to occupy the first position of the world, she could have caused a breach in the security shield of the Lightforce, which could very well have given the Dark Lord the opportunity to enslave the world with negativity and eternal bondage to the Desire to Receive for Oneself Alone.

When the Lord reminded *Nun* of her connection with *Tzadi*, *Nun* realized that her cosmic energy was inadequate for the protection of Earth's inhabitants. Cosmic *Nun* understood the Lord's reply and had no intention of jeopardizing His plans for the physical universe. All of the letter-energies were inspired by the subtlety and beauty of the natural world they were seeking to create.

With deep regret, the letter *Nun* departed from the presence of the Lord.

Destinies;
Cosmic Letters;
False Belief;
Physical Elements;
Natural Environments;
Paradigm;
Light of Wisdom;
Mundane World;
Reincarnation;
Perfection;
Monarchs;
Mercy;
Marriage;
Brides

CHAPTER 14

THE LETTERS
MEM AND *LAMED*

AND THE LORD SHALL BE
KING OVER ALL THE EARTH;
IN THAT DAY SHALL THE LORD
AND HIS NAME BE ONE.

—ZECHARIAH 14:9

The letters were eager to aid in the emergence of the infinite galaxies, the world, and man, but they had not yet discovered the method by which this future universe might be expressed. Perhaps they had at first imagined a small, orderly cosmos in which the dominant forms would be benign energy-intelligences such as themselves. In any event, their attempts to probe the deepest cosmological mysteries had failed, and each of their pleas had been denied.

Yet not only was the future of humanity at stake, but the future of the letters as well. The world they were about to bring into being would be their world too, and their destinies would be forever bound to it. Each letter symbolized a particular aspect of the all-embracing reality. Each differed from the others in many details, yet all were combined within a single ultimate reality. The letters rejected thus far had treated the universe as an entity independent of themselves. *Mem*, however, represented a new approach, one that sought to combine the channel for energy-intelligence with the energy itself, rather than viewing energy and channel as separate.

Mem מ approached the Lord and said, "Master of the Universe, may it please You to create the world through me. My energy-intelligence commences the word *Melech* מלך (King), which is Your Title."[258]

We are not generally aware in our everyday lives of the unity of all things. We tend to divide the universe into separate objects and events. Kabbalists tell us this sense of fragmentation came about when the twenty-two cosmic letters became physically expressed as the elements, the stars and planets, and the twelve signs of the zodiac. This is why we mistakenly perceive the universe as a hodgepodge of separate components. According to kabbalists, the false belief that all things are dissociated from each other is the essential reason for the growing disenchantment within modern societies. It is no wonder that life seems increasingly unreal, as contemporary man distances himself from his environment and even from his own self.

The positive and negative forces of Light and Vessel, and the energy-intelligences involving them, are truly not isolated entities but rather integrated parts of the whole. The real world cannot be broken down into independently existing units. Reality, according to the kabbalistic definition of that word, is endless and unified—everything, in other words, that this world seemingly is not.

Quantum theory has forced us to see the universe as a single web of thoughts and interactions. When we take a quantum-oriented look at medical science, we can no longer accept the commonly held notion that the individual elementary parts of our body are the fundamental reality. The unified whole consists of both mind and body. The external and internal aspects of the individual are woven into an inseparable net of mutually conditioned relationships. They cannot be treated as separate entities.

The most important characteristic of the kabbalistic worldview is the awareness of the unity and mutual interconnectedness of all entities and events. This was precisely the thought uppermost in the consciousness of cosmic *Mem*.

The idea of fragmentation governed Satan's vision of the universe. This concept lay at the root of a world enslaved by him. The Dark Lord sought to impose a framework of separateness upon the undifferentiated whole, leading human life toward psychic dismemberment and disintegration—a divide-and-conquer state of affairs that would allow him to establish total domination.

Cosmic *Mem*'s plea, then, represented her attempt to provide the energy-intelligence that would preserve reality as a unified whole. *Mem*'s plan was to endow the inhabitants of Earth with an awareness of the unity and interdependency of all seemingly separate manifestations, thus enabling mankind to grasp the essential meaning of existence and come to terms with the metaphysical principles that govern the whole universe.

Cosmic *Mem*'s plan took account of the fact that the free will, so intrinsic to the creation of the universe, at the same time gave the universe an inherent instability that severely limited its ability to sustain itself. Here, then, was the crux of the Creation dilemma. The alternative suggested by cosmic *Mem* would address this dilemma by leading mankind to a dramatically altered perception of reality.

Cosmic *Mem* exhibited the energy-intelligence of *Chesed* of *Zeir Anpin*. Her unique characteristics positioned *Mem* as a direct recipient of and channel for the awesome power of the Light of Wisdom. When *Zeir Anpin* succeeds in ascending to the level of consciousness of *Chochmah*, then the Light of Wisdom (which is referred to as Life), is revealed directly from the Countenance of the Lord into the universe. The mystery of *Chesed*'s peculiar position is revealed by King David in his coded verse: "The Lord will command his kindness (*Chesed*) in the daytime."[259] The implication, according to the *Zohar*, is that the Lightforce encased in *Chesed* is united with and manifested through every one of the *Sefirot* within the dominion of *Zeir Anpin*.

This passage is also reminiscent of the passage describing Day One of Creation: "In the beginning, the Lord created the Heavens and the Earth.... And the Lord called the light 'day' and the darkness He called 'night.' And the evening and the morning were one day.... And the Lord called the firmament Heaven. And the evening and the morning were the second day.... And the Lord said, let there be lights in the firmament of the Heaven to divide the day from the night."[260]

The first question that comes to mind when analyzing the above is: Why the need for so much repetition? Heaven was created on the Second Day; however, it is also referred to in the first verse, where the author states that it was created on the first day. Similarly, we are told that day and night were established on the first day, but the process seems to be repeated on the Fourth Day. Another strange feature of the Bible's description of Creation is its references to the "days." The first is called *Yom Echad*, meaning "Day One." The subsequent days are designated as the Second, Third, Fourth, and so on. However, the grammatically correct term for the starting point of Creation should have been the "First Day," not "Day One."

Before we proceed further, it is helpful to recall that the essence of Creation is the progressive evolution and transformation of the energy-intelligence of Vessels, which are known as the *Sefirot*. This process of transformation does not alter or affect the Lightforce or Light of Wisdom; it merely reveals a different aspect of the all-embracing, unified reality.

This process might be likened to water poured into a number of different-colored glasses, where the water might appear to be a different color in each glass, but the water itself is not altered or affected. The *Sefirot* or energy-intelligences operate similarly, diffusing and diversifying the original energy of the *Ein Sof* (Endless World).

On Day One of Creation, the energy-intelligence of *Chesed* was brought into being. This *Sefira* represents a unified entity. Like a seed, it contains within itself the full potential for all future growth and manifestation. This is why the Bible refers to the first day as *Yom Echad*, or "Day One," meaning a unified whole. The code word *echad* stresses this metaphysical undifferentiated state.

The coded names of the second and subsequent days indicate that they gave rise to multiform energy-intelligences. The word *sheini* (second) implies that there were now two forces and not the heretofore unified whole. *Chesed* is therefore the highest form in which the Lightforce—the Lord's all-embracing Light of Wisdom—manifests itself. Before the Second Day, the energy-intelligences were without any differentiation. They did not assume distinctive, personal configurations until after the emanation of the *Sefira* of *Gevurah* on the Second Day, when all the potentialities implied in every *Sefira* become manifest under the influence of its specific formative principle. In each subsequent *Sefira*, the Lightforce became apparent as a distinctive energy-intelligence.

This brings us to another question: Why is the differentiation between Heaven and Earth, and between day and night, mentioned as having taken place on *Yom Echad* as well as on the Second Day and Fourth Day? The *Sefira* of *Chesed* contained all these varied elements as a unified whole, like the seed that includes the root, trunk, branch, and leaf. On *Yom Echad*, the differing entities of Heaven and Earth were concealed as potentialities of the all-embracing whole. They did not appear as separate entities until further transformations took place, and these occurred on the Second Day when Heaven appeared as a distinctive cosmic force. On the Fourth Day of the cosmological process, the differentiation between day and night became manifested, whereas on Day One, day and night did not appear as separate entities.

Thus, it was a key attribute of cosmic *Mem* that her energy-intelligence composite was similar to the all-embracing unified Lightforce of the Lord. *Mem* is the total intelligence existing in the universe of *Zeir Anpin*, our universal system. She had a complete link with the Lightforce. The other *Sefirot* were not entirely connected. They could only be considered companions of the Lightforce. Consequently, *Mem* pleaded to the Lord, "Within me, You are called King. When the Lightforce of the Lord is revealed through me, the Dark Lord will no longer be capable of seizing Your Light. Thus, the Final Correction is assured by my energy-intelligence."[261]

The cosmic letter *Mem* and the Messianic element referred to in Rav Isaac Luria's (the Ari's) doctrine of *Gemar HaTikkun* (the Completion of the Correction) are welded together. The tikkun of the *Gemar HaTikkun*, the path to the end of all things in the World of Action, is also the path to the beginning. The emergence of all Essence from the Lord, as exemplified by *Mem*, becomes its opposite, the doctrine of *Gemar HaTikkun* (Final Correction) and the return to the original contact with the Lord, which resembles cosmic *Mem*.

The *Zohar* states that in the days of the Messiah, "There will no longer be the necessity for one to request of his neighbor, 'Teach me wisdom.'[262] One day, they will no longer teach every man his neighbor and every man his brother, saying 'Know the Lord.' For they shall all know Me, from the youngest to the oldest of them."[263]

Every event, every part of the continuum of existence, exists at once inwardly and outwardly. The Coming of the Messiah, or the *Gemar HaTikkun*, will mean that our mundane world has received its final shape. When man is in close contact with the attribute of sharing, the Lightforce of the Lord becomes manifest. When the activity of the individual—the Vessel's Desire to Receive—is associated to the Lightforce, the Desire to Impart is revealed. As with cosmic *Mem*, where Light and Vessel are one, it is man who adds the final touch to the Lord's Dominion. It is man who completes the enthronement of the Lightforce. In some spheres of being, Divine existence is intertwined with human existence. In a sense, therefore, we are masters of our own destiny. In the last reckoning, we, ourselves, are responsible for the continuation of our exile.

Mem saw herself as the ideal instrument for Creation because she alone personified the grand unification of the universe. She felt that if she were used as the channel for Creation, then the world would be assured of achieving the *Gemar HaTikkun*. Through the letter *Mem*, the whole of *Zeir Anpin* becomes crowned *Melech*, the King. *Mem* is the first letter of the word *melech*, and it is through her that the Lightforce becomes disclosed to the world. "Consequently," declared the *Mem*, "through me, the Final Correction of the world can be assured." The Lord's reply to *Mem* מ was strikingly different from His refusal of the other letters: "What you have said is assuredly so. But I cannot employ you in the creation of the world because the world requires a King. Return, therefore, to your place along with cosmic *Lamed* ל and cosmic Final *Kaf* ך . All three of you comprise the letters and cosmic influence of the word "king." The world cannot exist without a *Melech*, a King."

As strange as it may seem, cosmic *Mem* was indeed suited for the role of Creation. The Lord's reply did not negate her qualifications. On the contrary, the Lord acknowledged her great value and power—and those were the exact reasons why he did not want to use her for Creation. "The world could not exist without *Mem*'s participation," stated the Lord.[264]

The unity of the Lord and His Revelation, as exemplified by the King, was conceived as a relationship between Him and His Creation. This means that the Lord desires to have a relationship with something outside Himself, not only a relationship between Him and Himself. But if cosmic *Mem* was a living incarnation of the Lightforce, reflecting the hidden meaning and totality of existence, then why not make use of her cosmic channel for Creation?

The Lord told *Mem* that the Lightforce that would affect the perfection of the world needed to be encased in the three letters *Mem*, *Lamed*, and *Kaf*, which together constitute the word *Melech* (King). Were it not for this cosmic union, there would be no King, and without a King, the world could not exist. The essence of the Lord's reply to the letter *Mem* is that *Mem* does indeed symbolize the great *Chesed* mentioned in Psalms. She united with every *Sefira* in the body of *Zeir Anpin*. Her unique power is illustrated by the aperture within her thick base, indicating an emission of an abundant supply of the Lightforce.

Nevertheless, her cosmic power depends upon and coincides with cosmic *Lamed*. The letter *Lamed* revealed itself as "a tower flowing and flying in the air."[265] Cosmic *Lamed* towered high, reaching far above the others. Her rising structure connected the conscious level of *Binah* with the whole of *Zeir Anpin*. *Mem* was the reservoir and channel of the Lightforce for *Zeir Anpin*-consciousness; *Lamed* was her link.[266]

Kaf, as the final letter in the coded word *Melech* (King), was the *Malchut*, the connection, the final stage in which the creative Lightforce of the Lord materializes. The Lightforce cannot be manifested within a single metaphysical framework, but only through the passage from one period of time to another. Furthermore, there could not be a King without a Kingdom (*Malchut*). In these three ways, *Malchut* distinguishes herself as the illuminator of *Zeir Anpin*.

Malchut serves as the Celestial Throne of *Zeir Anpin*. Its essence is not absorbed through contemplation of the Lord's true nature, but rather by a perception of His appearance on the Throne, as described by Isaiah the Prophet: "The King is sitting upon a high and enhanced throne."[267] In the *Zohar*, the most frequently used illustration of the throne doctrine is the Throne of Judgment on *Rosh Hashanah*. "When Isaac—the *Sefira* of *Gevurah* (Judgment)—takes hold, then higher and lower angels assemble for judgment, and the Throne of Judgment is exalted. The King takes his seat thereon and judges worlds. Then is the time to 'blow the trumpet on the new moon at the appointed time on our solemn feast day.'"[268]

"Happy are the people of Israel who know how to remove the Throne of Judgment and set up the Throne of Mercy. How? With a Shofar."[269]

Malchut, the recipient of the Throne, becomes manifest in accord with man's activity. Thus, *Malchut* provides a certain representation of the Lord, but the specific characteristics of that representation depend on the Celestial Throne that humankind has erected through its actions. The Dark Lord hopes that mankind causes the Throne of Judgment to appear, for then the Dark Lord's long-awaited opportunity arrives, and the celestial wars begin.

The Dark Lord provided his fleet with sufficient striking power to accomplish its program of action, but the time would come when he would suffer a major setback that would break his back. The extinction of the Dark Lord was virtually assured at the time of the Exodus: "At that hour [the time of the splitting of the Red Sea], the commander appointed to represent Egypt in the Celestial Realm arrived, accompanied by six hundred chariots, directed by six hundred angelic adversaries of Israel. In order to face all the grades of fighting principalities and powers, all the hosts of celestial representatives of the enemy came. 'Where did they come from?' asks the *Zohar*. We have learned that the Dark Lord provided the commander of the Egyptian celestial army with the additional fleet to aid him."[270]

Moses was assigned the task of leading the children of Israel out of Egypt, where they had been slaves to the Egyptians. The biblical account shows Moses displaying awesome powers to perform ten miracles, as he called Ten Plagues down upon the Egyptians. The Bible does not go into much detail regarding how the Lord performed these miraculous feats, but one thing is certain: The control of celestial space was a crucial element of the Lord's plan to help Moses bring the Israelites out of Egypt. The battle for control of the mundane universe depended on the control of the Throne; or more precisely, it depended on which Throne the power of the Lord manifested upon.

The overthrow of governments, or the "thrones" of reigning monarchs, is determined by the outcome of a celestial battle being fought in the Upper Worlds. The victor of a celestial war subsequently becomes manifest on the level of our mundane planet. If the Egyptians were to be defeated and freedom was to reign throughout the land, the Egyptian celestial throne of judgment would have to be overturned or replaced by the Throne of Mercy of the Lord.[271]

Rav Shimon, the author of the *Zohar* stated, "It is now fitting to reveal mysteries connected with that which is Above and that which is Below. Why is it written here: 'Come (*bo*) unto Pharaoh?' Ought it not rather to have said: 'Go (*lech*) unto Pharaoh?'"[272] The verse seems to be saying that the Lord was instructing Moses about the next plague with which he was to threaten the Pharaoh; therefore, the instruction to Moses should have been phrased as a commandment to "go" to the Pharaoh, not to "come" to him. The *Zohar* says the word "come" indicates that the Lord summoned Moses to His Celestial Realm. In fact, the Lord guided Moses through a celestial labyrinth into the abode of a mighty supernal dragon— Egypt's celestial commander—from whom many lesser dragons emanated. But Moses was afraid to approach the Pharaoh because the Egyptian fleet of Darkness drew its power from supernal regions.

The Lord saw that Moses feared the Dark Lord. None of the Lord's forces had yet been able to overcome him. Moses proclaimed, "Behold I am against you, Pharaoh, King of Egypt, the great dragon, the Dark Lord."[273] The celestial throne of Egypt, representing the supreme degree of the Desire to Receive, had replaced the Throne of Mercy. Consequently, the terrestrial throne of Egypt reigned as no throne ever had before. Egypt did not rule over the world before Israel settled there. Egypt is described as a "house of slaves."[274] And this is the meaning of the phrase: "Now there arose a king,"[275] that is, the supernal Dark Lord of Egypt rose up in strength and gained predominance over the chieftains of the other nations.[276] The throne doctrine now should be clear and understandable. *Malchut*, the Throne, reveals the all-embracing reality. This revelation comes about as a result of man's behavior. The channel through which this energy is revealed is cosmic final *Kaf* ך .

Final *Kaf* ך provides the cosmic veil or curtain before the Throne, just as *Malchut* provided *Zeir Anpin* with a curtain to conceal the glory of the Lightforce. Simultaneously, the letter *Kaf* is the first channel for the Hebrew word *kisu'i* כיסוי, meaning "covering" and "concealment." The word *kisei* כסא (throne), is a derivation of the word *kisui*. These two words are intimately bound up with each other. Their code reveals the true nature of their essence: the Lord's omnipotence while maintaining its concealment. Similarly, man must become dressed before he can reveal himself to the world. It is a paradox, but one crucial to understanding our universe. For this reason, the bent image in the letter *Kaf* is a symbol of concealment.

Malchut is compared to a garment, or more precisely to what is known in kabbalistic terminology as *Levushei DeKadrutei* (Garments of Darkness). When the Kingdom of the Lord becomes revealed, *Zeir Anpin* removes this garment of Darkness and thrusts it upon the nations that embrace idolatry and engage in the worship of celestial constellations. The divestment of this cloak of Darkness is symbolized by the stretched-out final *Kaf*. As a result of *Kaf* disrobing, Israel and the righteous are then positioned to receive the all-embracing unity of the Light.[277]

All generations and all existence are woven into this curtain. He who sees it penetrates the very secret of Messianic Redemption. The final period of the Age of Aquarius is already a visible reality to those who can see. The prayer of the ancient kabbalists sums it all up: "When will we perceive what no eye has yet perceived? When will we hear of the Final Redemption?" The third and final degree of *Zeir Anpin*-consciousness revealed by *Malchut* is the mystery of *Malchut* herself crowning *Zeir Anpin*. The paradoxical nature of the Kingdom crowning the King has remained a central theme of the oldest mystical doctrines of Kabbalah.

The King's Throne, which embodies and exemplifies the whole of Creation, is dependent upon His Kingdom, *Malchut*. In this connection, there are two strange verses that deserve special attention. When King Solomon entered the paradise of mystical speculation, he said, "A woman of valor is the crowning glory of her husband."[278] The second verse says: "Go forth, ye daughters of Zion, and behold King Solomon, with the crown wherewith his mother crowned him in the day of his marriage and in the day of the gladness of his heart."[279] This verse expounds upon the idea of "sacred marriage," which plays a central role in the *Zohar* and is a foundational concept for all subsequent kabbalists. What took place within this *Zivuga DeKedusha* (Sacred Marriage) was the union of the King with his Spouse/Companion, also represented as the marriage of the two *Sefirot* of *Zeir Anpin* and *Malchut*.

There cannot be a king without a kingdom. The *Zohar*[280] said that Rav Shimon bar Yochai and his students attached special mystical significance to the festival of the Feast of Weeks (*Shavuot*). No holiday could more appropriately portray the mystical marriage than the Feast of Weeks, which commemorates the Revelation and the giving of the Bible on Mount Sinai.[281] To the kabbalist, this Covenant between the Lord and Israel is regarded as a Sacred Marriage.

On the night of *Shavuot*, the bride makes ready for marriage with the bridegroom. It was fitting that all those belonging to the palace of the bride should keep her company. All should participate through a festive ritual in the preparations for her marriage.[282]

Israel Najara, the poet of the Safed circle, wrote a poetic marriage contract, a mystical paraphrase of the marriage document (*ketubah*) prescribed by Jewish law.[283] The kabbalistic ritual of Shabbat, and especially of the eve of Shabbat, underwent a noteworthy transformation in connection with the idea of the cosmic marriage. On the basis of Rav Shimon's *Zoharic* concept of *Shabbat*, the kabbalists of Safed developed a ritual framework for achieving the highest level of consciousness possible in the terrestrial realm. The ritual's dominant theme was the mystical marriage between King and

Queen. A profound and deep interpretation of the Celestial Realm of the *Shabbat* Queen made it possible for any man or woman participating in the procession to effect a complete identification with the *Shabbat* Queen.

On Friday afternoon, shortly before the onset of *Shabbat*, the Safed kabbalists, usually dressed in white, went out into the open field to meet and sing mystical hymns to the cosmic bride—the *Shabbat* Queen. The most famous of these hymns was composed by Solomon Alkabetz, a member of Moses Cordovero's group in Safed. The hymn begins: "Go, my beloved, to meet the Bride. Let us receive the internal energy of *Shabbat*."[284] This hymn, today sung throughout the world, combines the cosmic energy of *Shabbat* with the Messianic hopes for the Redemption of *Malchut*. On Friday afternoon, Song of Songs, King Solomon's beautiful masterpiece that elucidates the indissoluble cosmic bond between the King and His Queen, is also intoned for the cosmic Bride.

The wide range of meaning contained in the cosmic marriage of *Zeir Anpin* and *Malchut* gave rise to the spiritual concept of soul mate,[285] and marriage at the terrestrial level assumed a profound significance. A spiritual union enhanced the cosmos, and at last, the people could feel at one with the Celestial Realm and the harmony inherent within the cosmos. Marriage and love between husband and wife thus became a central part of our cosmology. Human relationships would henceforth determine, to a large extent, the harmony of the universe.

Perhaps more importantly, with the advent of this symbolic union, a woman of valor could maintain and retain the universal harmony of our cosmos—no small task, even for a woman. If she could understand the significant role she was cast for on the stage of human existence, she would herself be crowned, as *Malchut* crowns the head of *Zeir Anpin*.

The verse in Song of Songs illustrates the importance of the *Kaf*. She is the first letter and initial channel of the energy-intelligence of *Keter*, the Hebrew word for "Crown." The importance of *Mem* has already been amply illustrated above; it was this very importance that led the Lord to command her not to descend below her majestic position in the cosmos, for her energy-intelligence was closely connected with the King.

And so *Mem*, head high, returned to her position in the Celestial Realm.

Internal Energy;
Revolutionary Action;
Mayin Nukvin;
New Age of Physics;
Newton's Universe;
Motivated Consciousness;
Feminine Polarity;
Mayin Duchrin;
Kedusha;
Doctrine of the Throne

CHAPTER 15

THE LETTER
KAF

כ/ך

DO YOU KNOW
THE ORDINANCES OF
THE HEAVENS?
CAN YOU ESTABLISH
THEIR ROLE ON EARTH?

—JOB 38:33

T he letters were searching for their place in the cosmos. This search compelled them to ask such questions as: *Who are we? What is our internal energy-intelligence? Where are we in this vast expanse of the cosmos?*

When cosmic *Mem* retreated from the stage of Creation along with cosmic *Lamed* and final *Kaf*, cosmic *Kaf* took a most radical, daring step, unprecedented in the Lord's Kingdom. At the moment of cosmic *Mem*'s departure, the letter *Kaf* descended from her Throne of Glory and returned to the arena of Creation. Trembling, she said, "Lord of the Universe, may it please You to commence the art of Creation through me. I am the first letter and initial energy-intelligence of Your *Kavod*, Your Honor." When cosmic *Kaf* committed her revolutionary action of leaving her Throne, the Throne trembled, and two hundred thousand galaxies were shaken to their very cores, bringing them to the brink of ruinous collapse.[286]

Similar catastrophic events are recorded in the Scriptures. "The Earth trembled, and the Heavens dropped,"[287] records the prophetess Deborah in the Song of Deborah. "The Earth shook, the Heavens also dropped at the presence of the Lord; Even Sinai was moved," declares the psalmist.[288] Various sources from many traditions describe catastrophic upheaval on Earth and celestial worlds in collision. Job says that the cosmic catastrophe at the end of a cosmic age is a Divine force "that removes the mountains, overturns them in His anger, shakes the Earth out of her place."[289] These natural upheavals took place all over the universe, some of them more severe and some of them less so. Cosmic *Kaf*'s descent from the Lord's Throne almost brought to an end the whole of celestial harmony in one sweep.

Most of us never will personally experience a natural upheaval more violent than a hurricane or an earthquake. Yet with a single action, cosmic *Kaf* may have achieved the dubious distinction of creating a disaster more severe than any hurricane or earthquake, a catastrophe affecting some two hundred thousand galaxies in their entirety. Such an event is unthinkable. Indeed, nothing of this magnitude ever has been recorded anywhere outside the *Zohar*.

The reappearance of *Kaf* in the creative process brings up several closely related problems that must be discussed. The first point is cosmic *Kaf*'s reason for ignoring the Lord's answer to her request. The Lord had told her that her significant role in the structure and maintenance of the Throne was the link that brought the Lord together with His Kingdom, and this was why she could not be the channel for Creation. The Lord could not spare her from the crucial job she was already doing for Him. We have examined the three fundamental principles that may be said to govern the cosmic position and general view of *Kaf*, but despite these reasons, cosmic *Kaf* insisted on pressing her daring suggestion that she be the channel for Creation. She did not exercise restraint. She was determined to carry her plea as far as possible.

The pleas the various letters were making to the Lord concerning their suitability for use in the Creation are more than just an interesting cosmic dialogue. In kabbalistic teachings, the power of thought is seen as the driving energy that causes the Lightforce, or ultimate reality, to become expressed.[290] This observation, however, carries us still deeper into the problem of the pleas themselves. The case each letter presented for her suitability to be the channel for the Creation of the world is similar to the labor involved in setting in motion the cosmic energy-intelligence of *Mayin Nukvin* (Female Waters), which manifests *Mayin Duchrin* (Male Waters). The expression of the cosmic Lightforce of the Lord—the all-embracing unity, *Mayin Duchrin*, which provides the essential life-energy-force—depends upon the characteristics of the channel through which the Lightforce is expressed. The letters are the vehicles for *Mayin Duchrin*. Each letter's plea was an expression of her consciousness, and it was that expression that created the structural reality of the letter.[291]

How can consciousness create its own Vessel? Is this merely some abstract theory that bears no relation to reality? Or does this concept have some basis in science? At first, the answer would seem to be no. The Newtonian universe with which we are all familiar was constructed from a set of basic building blocks—particles and forces in measurable quantities and with predictable effects. But the worldview suggested by modern physics indicates that this idea is no longer tenable. Scientists increasingly see the universe as a dynamic web of interconnected events, an indivisible whole in which all things participate. Everything in the universe is connected to everything else.[292] Even human consciousness participates in the cosmic whole, as both the new physics and the ancient Kabbalah tell us. In the words of renowned Kabbalist Rav Yehuda Ashlag: "The whole of the reality-structuring system of Creation owes its existence to thought-consciousness."[293]

The concepts prevailing in this new age of physics may seem strange to our Western sensibility at first. It is easy to become confused when confronted by phenomena that may exist beyond space and time. But more striking than all of these strange ideas is the fact that they are not new concepts. Indeed, anyone exploring kabbalistic material comes away with the unavoidable conclusion that much of the "new" information emerging today has been known for centuries. There are so many parallels between the teachings of the kabbalist and the teachings of the contemporary scientist that one almost finds it almost impossible to distinguish who said what.

The crucial implication of the pleas expressed by the letters to the Lord is that consciousness is not only a necessary part of the cosmic whole it is the essential component that controls and determines the structure of reality. This is a most important point in Kabbalah: The letters are motivated by consciousness. In fact, whether they do or do not exist, how they become manifest, the degree of their power, rests entirely on and is bound up with consciousness.

In Kabbalah, the universal interrelationship between reality and consciousness always includes the human observer and his consciousness. According to the kabbalists, *the plea itself is what creates*. The plea is the reality that lies behind physical matter. It encompasses all possible realities. In making their pleas, the letters presented a positive structure of reality.

In kabbalistic language, this plea is termed *Mayin Nukvin*, the female waters or Vessel that contains *Mayin Duchrin*, the male waters of the Lightforce of the Lord. *Mayin Nukvin* is the consciousness that originates within the Vessel, the receptacle. The *Zohar* says this feminine polarity determines the extent to which the Lightforce, the all-embracing unity, becomes manifest: "There can be no arousal from Above—*Mayin Duchrin*—unless there is an arousal from Below—*Mayin Nukvin*."[294]

Therefore, the response of the Lord to each letter of the twenty-two letters of *Zeir Anpin* and *Malchut* was to construct *Mayin Duchrin*, the internal consciousness of the Lightforce. The Lightforce began to manifest as the level of consciousness of each letter, this response essentially consisting of interplay between *Mayin Nukvin* and *Mayin Duchrin*. The plea and the response to the plea are not only active in the sense of determining the art of Creation; the letters themselves are the process.

Each letter's claim that she deemed herself an acceptable Vessel for the purpose of Creation is analogous to the process involved in the dynamic interplay of *Mayin Nukvin* and *Mayin Duchrin*. This interplay permits *Zeir Anpin* and *Malchut*, the channels for the manifestation of the all-embracing unity, to bestow Light upon our world. By utilizing a given letter's dimension of *Mayin Duchrin*, the Light is bestowed in a manner specific to and determined by that letter.

Each letter's plea triggered the Lightforce of the Lord to reveal its ruling power. The Lord's response to each letter was intrinsic to this revelation. At the same time, the revelation would disclose why each letter was deemed inadequate as a channel for Creation. The letter that would ultimately serve as the channel for Creation had to also serve effectively in the Lord's Kingdom and had to be able to withstand Satan. A letter's success or failure would depend entirely on her ability to contain the Lord's all-embracing reality. Her failure to withstand an onslaught by the Dark Lord would come as the direct result of the insufficiency of that letter's dimension of Light.

This was a conflict that the Master of the Universe had to win, and it was a conflict that could not be avoided. The ultimate objective of Creation was to provide human control over their individual destiny, and it is the struggle between Dark and Light that makes free will possible. In the kabbalistic view, there is no conflict between determinism and free will. To create free will, Satan pitted himself against the Lord's Structure of *Kedusha*. "The Lord created one against the other," declares King Solomon.[295]

King Solomon appears to offer humankind the unique ability to penetrate and influence the structural reality of the universe in a way undreamed of in the days of Newton. Furthermore, the centrality of human beings in the cosmic whole essentially demolishes the concept of causality and forces us to address the problem of sequence. The concept of universal time with an absolute past, present, and future comes into question. From a kabbalistic point of view, the possibility that an effect may precede its cause is something very real. "There is no past or future in the Bible. The historical presentation of the Bible is not one of sequence."[296]

The sequence of the *Sefirot* demonstrates an indeterministic universe. The *Sefira* of *Yesod* is portrayed in the Bible as Joseph, whose birth precedes in time the births of Moses, Aaron, and King David.[297] However, in its *Sefirot* sequence, *Yesod* follows *Netzach* and *Hod*, represented by Moses and Aaron, respectively.

As per King Solomon: An orderly cosmic structure—the objective of the Lord's dialogue with the letters—does not conflict with the concept of free will. All that the dialogue between the Lord and the letters accomplished was to provide for a universe that would prevent the Dark forces from seizing control. The Dark Lord's influence would be deeply felt, but never to the point of complete domination.

However, a letter incapable of containing the necessary dimensions of the Lightforce could not maintain a balanced, harmonious universe. The Lord, in pointing to the negative aspects of each letter, caused each one to depart and return to its place. Each letter's negative energy-intelligence was a wavelength that the Dark Lord could use to form a connection with that letter, seizing control of her in her entirety, thereby wresting control of the universe away from the Lord. Hence, the Lord compelled these letters to depart from the stage of Creation. Only a letter that was capable of maintaining a sufficient capacity of the Lord's energy-intelligence could resist an onslaught by Satan. Through the process of their dialogue with the Lord, the letters came to understand who should really be chosen as the proper Vessel for Creation of the world, and why.

The interconnectedness and interrelationship of the letters *Mem*, *Lamed*, and final *Kaf* in the word *Melech* (King) permitted the King—cosmic *Zeir Anpin*—to become revealed through cosmic *Malchut*. Thus, the Lightforce of the Lord became manifest within the universe. However, when the energy-intelligence of *Mem* became revealed—that is, when she entered her plea to the Lord—this interconnectedness and interrelationship ended. *Kaf* terminated her function as security shield and thus descended from the Throne of Glory. Her stretched-out version as final *Kaf*, linking the Upper and Lower Realms of the cosmos, thus came to an end.

This is precisely why the *Zohar* makes no mention of *Kaf*'s appearance and plea before the Lord, which it does for the other letters. *Kaf* was moved to approach the Lord only after the revelation of *Mem*'s energy-intelligence when *Mem* entered her plea. The interconnectedness among these three cosmic entities—*Mem*, *Lamed*, and final *Kaf*—was unique. They were strongly and permanently bound up with each other.

Consequently, when cosmic *Mem* became the dominating power within the cosmos, cosmic *Kaf*, too, was drawn away from her position at the Throne and descended, along with *Mem*, into the Lower Celestial Realm of our universe. Cosmic *Kaf*'s station, similar to that of the other letters, is located within the Cosmic Realm of *Briah* (Creation). There, individual and collective DNA became established and subsequently became physically expressed in the terrestrial and celestial realms of our universe. *Briah* provides the life-force for all the universes below.

"When cosmic *Kaf* stepped down from the Throne, she trembled, along with two hundred thousand galaxies." The binding link between the Celestial Realms of *Chochmah* and *Binah* and the Lower Worlds is *Malchut*. *Malchut* acts as the intermediary force in the cosmos. As the final *Sefira*, or cosmic force, in any creative frame of reference, *Malchut* bears the sole responsibility for any celestial or terrestrial entity becoming manifest. *Malchut*'s role might be compared to that of the seed of a fruit, which becomes physically expressed at the final stage of its development as the fruit. And in turn, that fruit bears a seed that is the *Keter* (Crown) or origin of the subsequent tree.

Cosmic *Kaf* maintained the continuity and interconnectedness between the two hundred thousand galaxies and the Lightforce.

The shape of the stretched-out final *Kaf* indicated and provided this cosmic bond. Thus, when *Kaf* left her position as *Malchut* of cosmic *Briah*, she caused a severance in that continuity as well as an interruption of the life-giving Lightforce to the Lower World of the two hundred thousand galaxies. The galaxies underwent a calamity of strangulation as the umbilical cord of the cosmos was threatened.

Elaborating on the mystical doctrine of the Throne, one of the most shrouded mysteries within the whole kabbalistic concept of the cosmos, Rav Ashlag provides the step-by-step description of the evolutionary process of the Celestial Throne:[298]

> *The Throne has three phases that encompass the whole array of the cosmic realm. The totality of the Lord's energy-intelligence, also referred to as the Lightforce, is based within the Upper Realm of the cosmos, known by the code name Atzilut. The Lightforce itself has a code name, Chochmah. When the Lightforce is transformed to the lower level of the Cosmic Realm of Briah, the Lightforce itself takes on a new coded designation: Binah.*

> *The upper section of the Throne contains six Sefirotic energy-intelligences, indicated by the four sides of the seat, along with the seat itself and the space above the seat. They represent the six cosmic channels of the Lower Worlds, known as Chesed, Gevurah, Tiferet, Netzach, Hod, and Yesod. The second aspect of the Throne is its four legs, which are the Mochin,*

which constitute the energy-intelligence itself, referred to as the brain. (This might be compared to the head of a man, wherein the brain, or energy, is stored. The body of man is the vehicle that allows the brain to become manifest). This source of all cosmic channels is known by the coded names Keter, Chochmah, Binah, and Da'at of the Lower Worlds. The third aspect of the Throne is Malchut of the immediate Upper Celestial Realm that descends to the next world below, establishing connection and continuity.

When cosmic *Kaf* descended from the Throne of Glory, the connection between the Celestial Realm of *Briah* and the Celestial Realm of *Atzilut* became severed. In the process, *Kaf* herself suffered the loss of her energy-intelligence, and "she, too, trembled." When cosmic *Kaf* experienced her loss at the Lord's battle station known as *Atzilut*, all subsequent connecting *Kaf* channels suffered this same deprivation.

Following this traumatic crisis, all two hundred thousand galaxies were in need of the energy-intelligence stemming from *Atzilut*. They trembled and were about to fall into ruins. The Worlds of *Briah*, *Yetzirah*, and *Asiyah* came close to annihilation at the hands of Satan because they did not contain sufficient Lightforce to withstand the Dark Lord's assault, which was honing in for the final kill.

Consequently, the descent of cosmic *Kaf* coincided with the Lord's reply: "Return to the Throne. If you were to leave the Throne permanently, *kilyah* (extermination) would be the lot of the Worlds. For this reason, you are the first letter and initiator of the cosmic force of destruction.[299] Your obligation to survive is owed not just to yourself, but also to the cosmos from which mankind shall ultimately spring."

Cosmic *Kaf*'s channel was too important in maintaining the harmony and interconnectedness of the vast cosmos. Thus, after the dialogue between the Lord and cosmic *Kaf*, *Kaf* departed to her significant position in the Lord's Throne of Glory.

Book of Formation;
Matter and Power of *Alef Bet*;
Antimatter;
Past, Present, and Future;
The Galaxies;
Midat HaDin and *Midat HaRachamim*;
Bread of Shame;
Interpretation of *Teshuvah*;
Am Segula;
The Milky Way;
Antimatter of Eden;
Atomic Harmony;
Age of Aquarius;
Dinosaurs Turning into Lizards

CHAPTER 16

THE LETTER
YUD

THERE IS NO EXCELLENCE
AMONG THE CREATURES
WHICH IS NOT TO BE FOUND
IN A MUCH HIGHER STYLE,
AND AS AN ARCHETYPE,
IN THE CREATOR;
AMONG CREATED BEINGS,
IT EXISTS ONLY IN
FOOTMARKS AND IMAGES.

—ALBERTUS MAGNUS

S tanding at the site of the Lord's dialogues with the letters and with a clear view of Creation, cosmic *Yud* speculated on what lay beyond. The plea of each letter injected a quantity of energy-intelligence of *Mayin Nukvin*, the Returning Light, into the cosmos.[300] These pulsating pockets of energy would create symmetry within the universe. Known in the lexicon of Kabbalah as the intelligent energy-force of restriction, the Central Column—Desire to Receive for the Sake of Imparting— became incorporated within the Lord's cosmic production.

In a modern sense, these pockets of energy-intelligence can be referred to as antimatter, which is the mirror image, the perfectly symmetrical counterpart, of ordinary matter. Matter is governed by the energy-intelligence, or thought-consciousness, of the Desire to Receive for Oneself Alone. The positive activity of future man in the terrestrial realm would trigger the energy-intelligence of cosmic *Mayin Nukvin*, and the response to this activity would be the Lord's energy-intelligence, *Mayin Duchrin*, the Lightforce, becoming manifest and revealed.

The problem, nevertheless, was to secure the existence and expression of *Mayin Nukvin* in the first place. This depended entirely upon humankind's activity, and from the looks of things, man was going to have a hard time keeping his head above water, let alone being concerned with the welfare of the cosmos. Little would he realize that his activities would create conditions affecting the entire universe.

Cosmic *Yud* and the other letters were fearful of the advantage Satan appeared to have over corporeal man and his physical universe. Any negative activity on the part of humankind played right into the hands of the Dark Lord. Egocentricity and selfishness were precisely the energy-intelligences required to maintain Satan. The Dark Lord played his cards well. Humankind's insatiable desire for momentary pleasure thwarted any counteroffensive that *Mayin Nukvin* and the Lord's fleet may have contemplated. So *Mayin Duchrin* retreated to the Lord's battle station, wondering whether mankind would ever activate sufficient *Mayin Nukvin* to repulse Satan.

The answer to *Mayin Duchrin*'s question lay in the remotest past of the cosmos—a past so ancient that the cosmos of today cannot even properly be said to be the same cosmos as the one that was first created. The scientist can shed no light on these matters from an epoch unimaginably distant from our own, but for the kabbalist, it is a simple matter to explore backward in time, using the *Book of Formation* and the *Zohar*. The scientist has not the slightest idea whether or not an infinitely older cosmos preceded our presently observed universe. Nor does he know how much antimatter there is in the universe or even how matter's counterpart came to exist. The kabbalist, however, is convinced that at one time, matter and antimatter existed as one. The *Zohar*[301] and the *Midrash*[302] postulate, with striking clarity, a scenario in which we have a two-stage cosmos. That is, for an infinite time, the universe behaved one way and then abruptly changed to behaving in a different way.

Before tackling the central question of whether the universe has altered its governing laws, it is important to assess what the *Zohar* has to say on the subject of primeval epochs:

> *In the beginning, before the energy-intelligences were created (long before the formation of the physical universes), the Lord first thought to*

create the intelligent energies with Midat
HaDin (Strict Judgment). However, further
interpenetration of these energy-intelligences
brought the realization that they had no chance
of survival once corporeal man made his
appearance. Hence, the Lord combined the
interconnected Midat HaRachamim (Compassion)
with Midat HaDin.

Rav Ashlag further explores the inner mysteries of Creation
when he asks:

Are we to assume that the Lord's thought
process is similar to ours? We mortals tend to
change our minds as events unfold and as we
gain further insight. The Lord's thinking process,
however, is not limited within sequential, linear
time. For corporeal man, there is a temptation to
assume that only the present really exists, but
the truth is past, present, and future are all
contained within the Source.[303]

Rabbi Ashlag continues:

However, when discussing the relationship of
cause and effect, as pertaining to emanated
beings, cause is expressed by "prior" and effect
by "subsequent." This is what the sages of the
Zohar and Midrash indicate when referring to
"prior" and "subsequent." The first cause or first
class of galaxies, known as the first universe,
was established and emanated with the energy-

intelligence of Midat HaDin, Strict Judgment of positive or negative activity. The Lord then brought the second stage of the universe into being. He added to the already existing energy-intelligences of negative and positive a third dimension energy-intelligence force of "compassion," known as the Third Column Force of the universe. There was no changing of the Lord's thought process. The second string of galaxies known as the second universe evolved through the natural law of cause and effect.[304]

Before exploring these sublime concepts of Creation, let us investigate the energy-intelligence systems controlling the two cosmoses. What is meant by *Midat HaDin* and *Midat HaRachamim*?

We can answer this question only when we realize that the universe has been programmed to evolve toward some final goal. The fact that our universe has been designed to fulfill an ultimate purpose is demonstrated by the striking compliance of material entities with natural laws. The Lord, the Cosmic Designer, arranged the world for the purpose of removing Bread of Shame,[305] which means that mankind plays a crucial role in the universal scheme. Yet today, man appears to play no role other than one of robotic consciousness. More than ever before, modern man sees himself as nothing but another component of a massive computer.

The Lord faced a grave, fundamental difficulty in fulfilling the purpose of Creation in the first incarnation of the universe, Cosmos I. This first version of the universe was based on *Midat*

HaDin (Strict Judgment). In such a universe, negative activity was followed by a consequent infusion of negative energy; positive activity triggered an infusion of positive energy. Strict judgment was the law that governed this cosmos, seemingly leaving no room for man to redesign the cosmic order.

Yet the operation of Bread of Shame meant that humankind surely could and would alter the cosmic order—but how? Within Cosmos I, automatic reaction to every action was inevitable. The results were always the same. Suppose an individual in Cosmos I committed an act of violence against his neighbor. The natural laws and principles of Cosmos I, based on Strict Judgment (*Midat HaDin*), dictated that the individual would be automatically punished for the crime of violence. If, on the other hand, a person chose activities of a positive nature, the Strict Judgment of Cosmos I dictated a positive reaction. All subsequent events or reactions in Cosmos I were determined by earlier decisions and actions, meaning that future events were entirely beyond anyone's control. The Lord saw this situation and understood there was a need for a cosmos that would permit the mind to act on matter, causing the mind to behave in apparent violation of the natural laws and principles of the universe.

Thus the Lord caused Cosmos I to naturally evolve into Cosmos II. Cosmos I consisted of two fundamental energy-intelligences: the Desire to Impart and the Desire to Receive, positive and negative, proton and electron. Cosmos II added a third dimension: the energy-intelligence of Restriction, the ability to nullify and temporarily force negative energy-intelligence to move in apparent violation of the laws of Cosmos I. Cosmos II allowed humankind to control its activities, offering it the ability to influence the structure of future events for the first time. The quality that allowed this new state of affairs to emerge was *Midat HaRachamim*, the energy-intelligence of Compassion.

Midat HaRachamim changed time from an absolute dimension to an elastic dimension. One way to understand time is to see it as merely a frame for the action of cause and effect. In Cosmos I, time was absolute, and cause and effect were strict. Action was immediately followed by reaction. Between the action and reaction, there was no opportunity or possibility for change. Cosmos II, however, allowed time to expand and contract according to context. Time was no longer absolute; it became relative, subject to the intelligent energy of *Midat HaRachamim* (Compassion). Thus was born the possibility of delayed reaction, a kind of suspension of time and causality that enables an individual to go back in time in a sense and nullify previous decisions and actions.

This, in essence, is the pragmatic interpretation of *teshuvah*, the word loosely translated as "repentance." However, this translation is a corruption of the cosmic code, for the literal translation of the Hebrew word *teshuvah*[306] is "return." The word "repentance" ultimately made its way into the lexicon of all religions following this corruption. In any event, from a kabbalistic point of view, a "return" in time to the past is a prerequisite for the nullification of negative activity.

Time is, in fact, a sequence of events, a distinct number of connected phases branching from each other in their order of cause and effect.[307] Our sensation of time is somehow more of a spiritual experience than a physical, bodily experience. Different people feel the passage of time differently depending on their situations. Two people working together in an office may experience the passage of an eight-hour day in completely different ways. For a person who loves her job, the day may fly by, while for someone who hates his job, the day may drag on interminably. This mutability of time is the mystery behind the energy-intelligence of *Midat HaRachamim*.

Let us now explore another aspect of Cosmos II by asking why the kabbalists chose the Hebrew word *rachamim* (compassion) to describe Cosmos II, which is represented by the power of the neutron to maintain balance and stability between the proton and electron.

Just what is this mysterious internal metaphysical force of the neutron? Time! Compassion! The demand for patience! The energy-intelligence that requests time and patience rather than strict judgment. In Cosmos I, negative activity resulted in an immediate negative reaction, but in Cosmos II, patience gives us a moment, a time of reprieve, the opportunity to set aside for a minute the laws and principles of strict judgment and display a little compassion. Time!

Unfortunately, we rarely display the virtue of compassion. When our fellow man hurts or wrongs us, our immediate reaction is one of strict judgment. Concern with settling the score or debt is uppermost in our minds. Seldom do we display the compassion that we ourselves demand of others. We usually judge without stopping to consider all that has taken place. We refuse to permit the "moment," Cosmos II, to be part of our cosmic order. We demand Cosmos II for ourselves when necessary, but we deem Cosmos I to be fitting and appropriate for others. Compassion is something that we require, but we do not feel that it is required of us.

However, the energy-intelligences of Cosmos I and Cosmos II have no double standards. The invocation of either Cosmos I or Cosmos II depends entirely upon the activity of humankind because humanity is the focal point of all cosmic activity. Man's display of compassion for his fellow man draws the cosmic order of Cosmos II into the universe. Similarly, man's demand for strict judgment—man's rejection of a moment of patience or compassion—infuses the cosmos with the law of Cosmos I, which holds no compassion for others, or for oneself. Consequently, we have no one to blame but ourselves for the chaotic condition of our universe. We are witnessing the lack of time in human activity.

Cosmos I and Cosmos II are both integral parts of this grand universal scenario, and in that fact lies the dilemma of the scientist. At present, human activity is unstable. It seesaws back and forth between Cosmos I and Cosmos II. Due to the enlightenment of the Age of Aquarius,[308] science has advanced to the point that we are now conscious of subatomic activity that seems to create uncertainty. Humankind is the most uncertain, least stable component of all Creation, and our actions exemplify this instability. Hence, humankind's scientific research can become befuddled rather than enlightened. But scientific achievement can also contribute to our enlightenment, at least insofar as our observations of the subatomic realm cause us to ask who or what the cause of nature's instability and uncertainty is.

The kabbalist asserts that humankind is behind all instability and furthermore that we have been provided with the antidote to this cosmic upheaval. Time! Compassion! True, this antidote may be a bit difficult to come by, but it is nevertheless attainable, and an enormous infusion of a time for compassion within the cosmos will ensure the dawning of the New Age.

This, then, was the concern of cosmic *Yud*: the assurance that human activity would overwhelmingly flood the Lord's cosmic realm with compassion, for this was the energy-intelligence that could overcome the threat of the Dark Lord once and for all. Was there a cosmic flaw inherent within Creation? The scientific community claims that mass extinctions (most notably the one in which the dinosaurs perished) have repeatedly punctuated the history of life on this planet. These cataclysms seem to occur with striking precision and regularity, a fact that has given rise to a spectacular new theory: There may exist somewhere out in space a companion to our sun, a star that has been given the ominous name "Nemesis." Galvanized by this radical proposal, astrophysicists have been searching the sky.

From the kabbalistic point of view, the cosmos was created in perfect harmony. The Creation account as described in Genesis allowed for no cosmic flaw. The extinction theories, however, seem to point to a world not exactly cast in the mold of the Garden of Eden. Even Noah's Flood could not be considered a just cause for the disappearance of the dinosaurs from the Earth. Yet despite all the information pointing to a history of catastrophic events and mass extinctions, we still have not the slightest clue as to what or who is behind it all.

> And the Lord said, "Let there be a firmament in the midst of the waters."[309] This is an allusion to the separation of the Upper from the Lower (immaterial from the material) through that negative aspect of existence which is known to

kabbalists as the Left Column. Up until now, Day One, the text of Genesis has referenced only the positive Right Column, but here it alludes to the Left, indicating an increase of discord between the Left and the Right.[310] It is the nature of the Right to harmonize the whole, and therefore the whole is written with the Right since it is the source of harmony. But when the Left Column awoke, discord was awakened, and through that discord the wrathful fire was reinforced and there emerged from it the Gehinom (Hell), which originated from the Left Column and continues there.

Moses, in his wisdom, pondered over this and drew a lesson from the work of Creation. In the work of Creation, there was an antagonism of the Left against the Right, and the division between them allowed the Dark Lord to emerge and to fasten himself to the Left. Thankfully, the Central Column, the mediating principle, intervened on the Third Day and allayed the discord between the two sides, so that Gehinom and the Dark Lord descended below. Thus the Left became absorbed in the Right and peace was restored."[311]

What seems to emerge from the *Zohar* is the establishment of the Dark Lord. However, to maintain harmony and stability within the Cosmos, the Third Day—or Central Column—became an integral energy-intelligence force in the universe, allowing the all-embracing reality to find expression as a unified whole.

Says the *Zohar*, "Hence we find twice written in the account of the Third Day 'that it was good.'[312] This day became the intermediary between the two opposing sides and removed discord. It said to this side 'good,' and to the other side 'good,' and reconciled the two. Connected with this day is the secret of the name of the four letters [Tetragrammaton] engraved and inscribed."[313]

Thus, the primordial state of the cosmos seems to be without flaws. The universe emerged in a state of order, coherent and organized. Our universe, from a kabbalistic point of view, was carefully fashioned and is extraordinarily uniform. Our existence, along with that of the cosmos, did not result from a colossal, meaningless cosmic accident. The cosmos displays coherence throughout. *Rachamim* ruled supreme as the supernatural Lightforce of the Lord's Empire, with antimatter as its energy-intelligence.

In the scientific worldview, however, the cosmos seems to have developed a preference for matter (the energy-intelligence of the Desire to Receive for Oneself Alone). Chaos appears to dominate our terrestrial realm. The energy-intelligence of Satan seems to have taken over the universe, lock, stock, and barrel. All one has to do is to observe the perpetual conflict and confusion all around us to see that the world is in disarray.

If there is cosmic design and order, set principles that play a role in physics and have the same *gematria* (numerological value) everywhere in the universe, where is the evidence of it? Where is the antimatter that would prove nature's symmetry? Where is the *Mayin Nukvin* King Solomon assured us exists in the universe? "The Lord created one against the other," declares Ecclesiastes.[314] Our explorations of the subatomic world have revealed the dynamic nature of energy and matter, allowing us to observe the interplay of the creation and destruction of particles within the cosmic web. Yet how is it that we, and the things around us, appear to be made up of only protons, electrons, and neutrons—without a trace of antimatter?

Cosmic *Yud* raised another set of questions for the purpose of understanding and coping with the chaotic conditions that humankind would ultimately face: "Where is the Dark Lord located? Why can't we observe him? What is his activity? How can we detect his energy-intelligence?"

Satan hides behind cosmic explosions, close to collapsing and colliding stars, and within the rest of the cosmic chaos that rocks the Celestial Realm. It is the insidious Satan who causes the magnetosphere to go into sudden convulsions that dump millions of amperes of current into the polar region, giving rise to the aurora borealis.

"Observe," says the *Zohar*, "that when the days of a man are firmly established in the supernal celestial, extraterrestrial grades, then man has a permanent place in the world. However, if man has not taken his rightful place in the cosmos of the outer-space connection,[315] his days descend until they reach the cosmic level where the Dark Lord resides. The Angel of Death then receives authority to take away man's soul and pollute his body, which remains permanently part of the dark side. Happy are the righteous who have not polluted themselves and in whom no pollution has remained."[316]

"The Dark Lord," further declares the *Zohar*, "is based within the Milky Way." This is indeed a dramatic revelation. For most observers of the heavens, the Milky Way is an unexplored celestial continent filled with exotic beauty and serene stellar entities. Yet within that breathtaking view of an astronomer's paradise is woven the nemesis of mankind. The *Zohar* makes it quite clear how our existence and our character are determined by the deep connection between man and the cosmos.

The *Zohar* continues: "Come and see: In the center of the galaxy, there is a celestial Dark Lord known as the Milky Way. All celestial entities, an endless and infinite array of them, revolve around it and are charged to keep watch over the secret deeds of human beings. In the same way, myriad emissaries go forth from the primeval celestial Dark Lord, the same Dark Lord by whom Adam was seduced. The Dark Lord goes forth to spy out the secret deeds of mankind."[317]

Thus, we see that human concerns are inextricably connected with the cosmos. And not just human concerns; the mass extinctions of history also have a cosmic cause. It is not important whether these devastating calamities took place in precisely the manner set forth by scientists. The question is: Why did they happen?

Most scientists favor extraterrestrial causes for these mass extinctions, despite the lack of evidence for such events occurring on a regular basis. The kabbalistic worldview cites the Milky Way as the villain. "The *klipot* of the Dark Lord are, without a doubt, dependent on man's action," states the *Zohar*. The crowded Milky Way galaxy is a floating island comprising more than a hundred billion stars, beautiful from afar, but also a terrifying force of potential disaster. The Dark forces are obviously capable of raining destruction upon any target they choose.

And who is the Dark Lord?

Again, we turn to the *Zohar*, which reads as follows:

> *Rav Chiya discoursed on the verse: "You have beset me behind and before and have laid Your hand upon me."*[318] *He said, "How greatly is it incumbent on the children of men to glorify the Lord! For when He created the world, He looked on man and designed to make him to rule over all earthly things. He was of a dual form, righteous and dark, and resembled both celestial and earthly things. The Lord sent him down in Splendor, so that when the lesser creatures beheld the glory of his state, they fell down before him in awe, as it says: 'And the fear of you and the dread of you shall be upon every beast of the Earth and upon every fowl of the Earth."*[319]

> *Rav Chiya continued: "The Lord brought him into the Garden of His own planting, so that he might guard it and have endless joy and delight. Then the Lord gave him the commandment concerning the tree.*[320] *And alas! Man failed in his obedience. Had Adam been obedient, he would have so dwelt forever, having eternal life and perpetual joy in the glory of the Garden."*[321]

Had Adam never sinned, the Dark Lord never could have raised his ugly head. The galaxy and the Earth would have been lovely places. The fall of Adam changed everything. His deviation from the path of virtue left the entire cosmos in a state of distress and diminution. The Earth, the galaxies, and the universe beyond changed from a scene of serenity and quiescence to one of chaos and catastrophe. The universe became anything but a Garden of Eden. The spiritual, Celestial Realm suffered this debasement no less than the terrestrial, physical universe. The *Zohar* puts it this way: "According to tradition, the fleshy part of Adam's heel outshone the orb of the sun. Rav Elazar said, 'After Adam sinned, his beauty was diminished and his height was reduced to a hundred cubits. However, before the Fall, Adam's height reached to the first firmament.'"[322]

Despite all scientific evidence to the contrary, the *Zohar* tells us that dinosaurs and other primeval life did *not* become extinct. They still exist now as they did before. But how can this be? Ancient writings tell us that after the Fall of Adam, the universe from time to time would experience a resurgence of the serenity and perfection of the Garden of Eden. The *Zohar* tells us: "It is written: 'King Solomon made him a palanquin *apiryon* of the trees of Lebanon.'"[323] In this quote from the *Zohar*, "*apiryon*" symbolizes the Palace Below, which is formed in the likeness of the Palace Above, which the Lord called the Garden of Eden.

In addition, the *Zohar* tells us that during the Temple period, the universe returned to its former cosmic position.[324] The Earth experienced an era of peace and a degree of tranquility surpassed only by that which Adam knew in the Garden of Eden. Human activity was impressed with *Mayin Nukvin*, the antimatter of Eden. We can therefore ascertain that the cosmos, while undergoing apparent change, maintains a parallel original universe that never changes. The alteration of the cosmos, as we mortals perceive it, happens solely on a physical, material level. Intrinsically, on the metaphysical level, nothing changes.

As Adam went, so went the universe. When Adam's corporeal framework declined in size and glory after the Fall (sin of Adam), so did all other species of plant and animal life. The physical universe with all its infinite galaxies underwent the same kind of contraction. The dinosaur shrank to the size of a lizard, and the pterodactyl became a bird! Plants and insects became like little cousins of their much larger former selves. The real universe, however—the metaphysical universe—did not undergo any form of extinction, mutation, or change.

On the temporal level, however, change does occur, including the periodic mass extinctions that science speaks of. But if the human realm is inextricably bound to the cosmos, then what is the cosmic agent that initiates these apparently regular celestial bombardments that wipe out life on the Earth? The *Zohar* strongly implicates man as the mechanism behind all cosmic activity. Thus, the Dark Lord of the Milky Way rains death and destruction in response to negative human behavior. The energy-intelligence of negative human behavior consists principally of the Desire to Receive for Oneself Alone, as exemplified by the sin of Adam. When humans behave negatively and give in to the Desire to Receive for Oneself Alone, the resulting infusion of negative energy into the cosmos fuels Satan.

Materially, the cosmos is asymmetrical; metaphysically, the universe is symmetrical in every way. The metaphysical universe is the absolute picture of perfection. Predictable and highly ordered, it is the Garden of Eden that soon will return to its original pristine splendor. Paradoxically, the New Age is also responsible for the success of high technology in man's search for infinite improvements and refinements of material processes, such as the development of fiber optics and the splitting of atoms. At the same time, wherever we set our eyes—from far-flung galaxies to the innermost recesses of the atom—we discover uniformity and an elaborate organizational structure.

It should follow, therefore, that our universe conforms to the beauty and elegance of atomic harmony. However, everywhere we look, we see evidence of an irregular, chaotic universe manifested cosmically by celestial bombardment on the earthly level by man's inhumanity to his fellow man. How can these two seemingly opposing forces—order and chaos—possibly be harmonized?

The cosmos, connected with the celestial Age of Aquarius, reveals the wonder of nature's intrinsic beauty and order. Spiritual man frequently experiences his cosmic connection with the all-embracing unified reality. Materialistic, self-centered, ego-minded individuals are the true culprits. They refuse to let go of the illusion that gives preference to cosmic disunity. *Mayin Nukvin*, compassion, is not in their spiritual lexicon. So they continue, without knowing that it is their negative activity that fuels the machinery of the Dark Lord. We humans are responsible for the chaotic universe, just as the blame may be laid on Adam for the transformation of dinosaurs into lizards!

Yet somehow, the physical universe maintains its essential symmetry. Otherwise, the planets' orbits around the sun would keep changing as the solar system rotates around the Milky Way. How can this be?

"And the righteous are the world's foundation," declares the Book of Proverbs.[325] Earth and the galaxies are made of ordinary matter, but from the *Zoharic* point of view, it is *Mayin Nukvin*, the antimatter energy-force of compassion, that maintains stability within the cosmos.

Says the *Zohar*: "Rav Isaac once asked Rav Shimon to explain how it is that some say the world is founded on seven pillars and some on one pillar, to wit, the *tzadik*, the righteous. Rav Shimon replied, It is all the same. There are seven, but among these is one called *tzadik* on which the rest are supported. Hence it is written: The Righteous One, *Tzadik*, is the foundation of the world."[326]

The kabbalists knew that the cosmos originated with both positive and negative energy, the former a cause and the latter its effect. This union of opposites was the realm of Cosmos I. However, when Cosmos II came into existence—once the mediating principle or Central Column became established—humankind no longer remained merely a participator within the cosmos; it became a determinant. Therefore, it also became necessary, for the continued existence of the universe, that each generation would have a permanent team of *Tzadikim* (Righteous Ones) to prevent the annihilation of the cosmos by the Dark Lord. These Righteous Ones maintain a sufficient reserve of *Mayin Nukvin* to assure a spatially symmetrical universe. The rest was left to humankind, either to fulfill the purpose of Creation and thereby vaporize the *klipot*, or to succumb to the energy-intelligence of the Desire to Receive for Oneself Alone and permit the Dark Lord's empire to rule over the entire cosmos.

The difficulty in overcoming the Dark Lord remained. Humankind needed a special cosmic support that would provide Earth's inhabitants with a fighting chance. Cosmic *Yud* believed that she had the specific energy-intelligence capable of waging a successful war with the *klipot* of the Dark Lord. The Tetragrammaton,[327] הּ·וֹ·הּ·יֹ the four-letter code name of the Lord, was her secret weapon. The Tetragrammaton contains the highest, purest, most concentrated form of energy imaginable. Indeed, it was by means of this cosmic channel that Moses slew the Egyptian.[328]

Rav Shimon continues: "We have a tradition that when the Lord created the world, He engraved into the midst of the mysterious, ineffable, and most glorious lights, the letters *Yud* and *Hei*, *Vav*, and *Hei*. They are in themselves the synthesis of all worlds both Above and Below. The Upper was brought to completion by the cosmic influence of the energy-intelligence of the letter *Yud* embodying the primordial celestial point that issued from the perfectly concealed and unknowable Infinite— *Ein Sof*."

As the *Zohar* tells us: "Out of the imperceptible Endless issued a slender thread of Light, which was itself concealed and invisible, but which yet contained all other Lights. The Light which came forth from the slender Light is mighty and frightful."[329]

The *Zohar* adds a strong holistic flavor to the quantum aspects of the nature of energy with the assertion that everything is made of everything else. Yet there remains a hierarchy within the cosmic structure. It is within the all-embracing unity of cosmic *Yud* that the constituents of energy and matter became the ultimate unified force. Thus, cosmic *Yud* contended that her dimension of Light, combined with her energy-intelligence of the grand unification in the creative process of the world, assuredly would bring about the Final Correction to the world.

The energy emitted by *Yud*'s plea, as stated before, established the energy-intelligence of *Mayin Nukvin*. Here was the opportunity the Dark Lord had been waiting for but had never thought would become a reality: the possibility of making contact with *Yud*'s intelligent energy.

The Lord's response to *Yud*'s plea was fast in coming: "Stop right where you are!"

By extending herself beyond the protection of the cosmic shield, *Yud* would provide the Dark Lord with an advantage that would allow him to rule over cosmic space, the world of good and evil, and the universe of corruption. Moreover, mere contact with the Dark Lord would prevent cosmic *Yud* from regaining her place in the Tetragrammaton. The World of Action, Earth's cosmic universe, is governed by change—from correction before the sin of Adam to corruption and once again back to correction.

For this reason, the Tetragrammaton is pronounced "*Adonai*"[330] and not given its original pronunciation. When positive human activity elevates the universe to its true connection with outer space and thus to its completion, the Tetragrammaton will once more be pronounced as it is written. This cosmic phenomenon will only come to be recognized upon completion of the process of *tikkun* (correction).

Thus, cosmic *Yud* was told to keep her place. Were she to be used in the creation of the world, she would then be at risk of becoming corrupted. This would have the effect of rooting out *Yud* from the Holy Name, and in the Holy Name, no corruption ever takes place. This cosmic process is revealed in the words of the prophet Malachi: "For I, the Lord, change not."[331]

Changes in celestial entities, the dynamic interplay of cosmic forces, unprecedented supernovae, dinosaurs turning into lizards—these are phenomena that exist in the unreal, illusionary cosmic dimension of the World of Action, a universe of which man is an integral part. Inasmuch as man himself is fragmented, defects, extraterrestrial bombardment, and corruption are aspects of this world's normal existence. Technological advances help us to discover many "new" cosmic features, yet the discovery of each new phenomenon relates to the cosmos of change, the unreal rather than the all-embracing unified reality. Indeed, the World of Action, Earth's cosmic universe, is so governed by change—from correction before the sin of Adam to corruption and once again back to correction—that it was no place for cosmic *Yud*. Thus, the Lord said to her, "You are engraved within Me, marked within Me, and My Desire and energy-intelligence is in you. Consequently, you are not the suitable cosmic channel for Creation."[332]

CHAPTER 17

THE LETTERS
TET AND *CHET*

BEN AZZAI SAID,
"THE REWARD OF A GOOD
DEED IS ANOTHER GOOD
DEED, AND THE REWARD
OF TRANSGRESSION IS
ANOTHER TRANSGRESSION."

—*PIRKEI AVOT* (ETHICS OF THE FATHERS)

The face of the Earth, the face of our solar system, the view of our galaxy and of the infinite universe beyond— all have undergone numerous changes caused by human activity in them. They have all been affected by the physical uncertainty that characterizes the realm of changing illusions. But with planets in their permanent orbits, satellites rotating with clockwork precision, and seasons coming uninterrupted in their order, humankind has had little reason to believe it had any hand in the larger cosmic scenario of colliding galaxies, celestial barrages, and comet strikes. For most of human existence, the Lord provided the necessary protection against almost every cosmic cause for alarm. Cosmic rays were kept well in hand by a magnetic shield. Destructive radiation was restrained by the ionosphere. Earth seemed assured of the right measures of heat and light, sufficient water, and the correct atmosphere to support life.

In the past couple of centuries, this has changed due to the rising tide of humanity and our reliance on science to solve all our problems. But despite the truly dangerous games that man has been playing by splitting atoms and genes, the real threat to humankind's existence is Satan, who continues unabated in his pursuit of dominion over the universe. To prevent this domination from coming to pass, cosmic *Tet* approached the Lord to enter her plea to be the celestial channel for the creation of the world. Cosmic *Yud*, for all her positive energy-intelligence, was needed at the *Gemar HaTikkun*, the Final Correction. Her position within the Tetragrammaton could, under no circumstances, be placed in jeopardy.

"Lord of the Universe," cosmic *Tet* said, "may it please You to create the world with me, since through me You are called *Tov*

(Good) and Upright. The initial letter of the word *tov* is *Tet*, indicating that my internal energy-intelligence is good, proper, and positive."[333]

Cosmic *Tet*'s particular energy-intelligence was precisely the panacea needed to alleviate every danger and onslaught originating from Satan. In the optimal environment provided by *Tet*, humankind might very well achieve its objective: the removal of negative energy. Provided with and fueled by sufficient streams of positive energy, humanity would be prepared to begin the great battle to restore harmony in the cosmos. A return to the cosmic order of the Garden of Eden, the pre-fall days of Adam's youth, seemed certain with cosmic *Tet* providing the blueprint for a new era.

According to the *Zohar*: "'And the Lord saw the light, that it was good.'[334] Every dream that contains the term *tov* presages peace Above and Below, provided the letters are seen in their proper order. These letters *Tet* ט, *Vav* ו , and *Bet* ב afterwards were combined to signify the Righteous One (*Tzadik*) of the world. As it is written: 'Say of the righteous one that he is good,' because the Supernal radiance is contained therein."[335]

In the foregoing chapters, we have discovered within kabbalistic teachings some radical ideas about Creation, astral influences, space and time, order and chaos, a meaningful universe, and an understanding of humankind and its cosmic environment. Much of what has been presented will undoubtedly raise some eyebrows and draw its share of criticism. Many of the readers of this book will no doubt reject outright some of the ideas presented by the kabbalists. However, such a response is par for the course when any new understanding of Creation, the universe, and man's role in this is advanced. Nevertheless, the kabbalistic view of the universe presents ideas that cannot be ignored, especially in light of recent advances in quantum physics.

Quantum theory permits the possibility that events in the subatomic realm could occur without a cause. Particles seem to appear out of nowhere and for no apparent reason. Discussing this scenario, physicist Alan Guth remarked, "It is often said that there is no such thing as a free lunch. The universe, however, is a free lunch."[336] I personally consider quantum mechanics to be a direct validation of the kabbalistic understanding of existence, in that quantum theory breaks the centuries-old deadlock between religion and science. In the subatomic realm, the laws of Newtonian physics are simply invalid and new principles take over, principles that allow for events of the kind that are described in the Bible and the *Zohar*. The modern worldview holds that religion is a myth, yet quantum theory teaches us that science is a myth also.

The possibility that space-time could arise without cause out of nothingness seems to contradict everything we know, despite the fact that quantum theory points to just such a finding. Using the kabbalistic dictum that there is sense to what makes sense, we can attain a different, more realistic perspective on these matters—a perspective that will perhaps prove to be more valid than that of the scientists themselves. If some entity apparently appears where none existed before, do not assume it has appeared without specific causation. When considering these phenomena, you must drown your ego and leave ignorance on the doorstep where it belongs. An overly rational approach to such questions only reflects the fact that our everyday consciousness, consisting mainly of input from our five senses, is wholly incapable of perceiving the metaphysical basis of reality.[337] From a kabbalistic point of view, the origin of the universe and the laws of nature evolve from and revolve around a basic cause: the removal of Bread of Shame.

The Age of Aquarius is indeed an age of enlightenment. Quantum theory forces us to ask questions about causality that themselves raise a host of issues—the very same issues that Kabbalah has been addressing for centuries: Can we be responsible for our acts in a world where events occur without cause? Are good and evil real? What are the underlying causes of existence? If there is a Creator, why does He remain beyond our perception? No longer can those who once worshipped at the altar of the great gods of Science and Technology avoid these questions and dismiss them as meaningless. As our scientific understanding of the cosmos advances, science inadvertently provides the conceptual framework necessary to improve our understanding of Kabbalah and of the cosmic code contained within the Bible.

Since the advent of Newtonian classical physics, a dichotomy has existed between religion and science. For the kabbalist, however, no such conflict ever existed. Like science, religion is also a myth. Only the Creator is real. The Bible is not a religious document of doctrine. Buried beneath the Bible's literal meaning is the answer to the secret of life's origin, the true nature of which it is the purpose of Kabbalah to decipher. Central to the truth is the idea of a beautiful cosmos of harmony and symmetry, with humankind playing the role of its determinator. The entire physical universe is the medium for the expression of the desire of humanity.

What does the Bible really mean when it says the Lord caused Creation to come into existence? How is Creation affected by the role of man as determinator? The word "creation" carries with it a variety of meanings. From a kabbalistic point of view, the creation of the universe is taken to mean the creation of energy-intelligences to provide free will, as exemplified by the Desire to Receive with the opportunity of removing Bread of Shame.[338] The creation of the observable physical world, including space and time, permitted humankind to corporeally express the Desire to Receive, making it possible for humanity to achieve this objective.

The Creator's withdrawal and restriction (*Tzimtzum*) were, consequently, necessary prerequisites of Creation. The Lord was then and remains now the true Composer of all Creation. However, it was impossible for anyone who dwelt in His presence to experience the Desire to Receive, so effulgently did His Lightforce flow out upon those near Him. Therefore, the Creator restricted Himself to allow His creations—the souls of man—the opportunity to remove Bread of Shame. Put another way, He changed His role from composer to arranger. Henceforth, after the Creation, the music of the universe would depend upon the behavior of humankind.

The biblical code contains the natural laws and principles for an orderly universe. The essential feature in all of these laws is the removal of Bread of Shame. Through his own actions, man can open the gates of his own being, revealing a universe of endless beauty and perfect harmony. The old idea of man's purpose in the universe, where his only function was to "serve the Lord," has been swept away by the concept of man's capacity for self-determination.

Armed with this information, we can now embark upon a simplified explanation of the difference between good and evil. Good activities are those that take into account the purpose of Creation, namely the removal of Bread of Shame. Evil, on the other hand, takes no heed of the purpose of Creation and does not seek to remove Bread of Shame.

However, not all negative energy-intelligences are necessarily evil. The Desire to Receive for the Sake of Imparting, while negative in the sense of cosmic polarities, still conforms to the natural laws of the universe. The negative inclination is a necessary factor in the continued existence of the world,[339] for without it, no one would build a house, marry, raise a family, or engage in trade. Nevertheless, it is within humanity's ability to control negative energy-intelligences, against whose power the cosmic code of the Bible is an antidote.[340] This control permits humankind to make manifest a physically expressed universe in harmony with the all-embracing unified reality of the Lord.

Perhaps the most widely accepted explanation of suffering is that people are undergoing punishment for their sins. If they join the righteous, then they can enjoy their full reward in the World to Come. The term "World to Come" poses a number of important theological questions that sages have attempted to solve in a variety of ways. First, there is the issue of where this world may be. Another vexing problem is why we must wait until after death to merit entrance to the World to Come. Some scholars have stated that this entire issue is beyond the grasp of man's intellect. The verse "The secret things belong unto the Lord"[341] is often quoted in support of this position. Others have proposed a variety of partial solutions. From a kabbalistic point of view, however, the World to Come exists right now for those who can access it; it's a place where cosmic consciousness begins, and where the limitations of time, space, and motion have no dominion.

The Lord viewed the cosmos as it emerged, and said, "It is good." Before the sin of Adam, negative energy-intelligence did not exist, and there was no violation of the doctrine of Bread of Shame. When man adheres to the laws and principles of the world of reality, he does not expect a free lunch. Human activity that stays clear of the free-lunch concept ensures that one is a *Ben Olam Habah*, a Son of the World to Come. The individual for whom the concept of Bread of Shame means nothing is hitching a ride to nowhere on transportation provided by the physical world of illusion.

"The *Tzadik* will be rewarded with material prosperity, and his merit will endure forever."[342] This is how the *Zohar* describes the righteous as individuals whose behavior forged a cosmic connection to the energy-intelligence of *Zeir Anpin*, or the *Sefirotic* consciousness of *Yesod* (Foundation).[343] Entering this realm may be compared to what happens when a spacecraft escapes the Earth's gravitational field: Upon entering outer space, it becomes governed by a new set of laws and principles regarding time, space, and motion.[344]

Again, we turn to the *Zohar*, where we see Joseph refusing to succumb to the temptation of a free lunch: "And it came to pass after these things that Joseph resisted his master's wife…"[345]

Rav Chiya discussed the text: "Bless the Lord, you angels of His, you who are mighty in strength, that fulfill His word, hearkening unto the Voice of His word."[346] Rav Chiya interpreted this to mean: How greatly it behooves a man to guard against sin and to pursue the straight path, so that the Dark Lord, his daily assailant, should not lead him astray. And since he assails man perpetually, it behooves man to muster all his force against him and to entrench himself in the place of strength; for as the Dark Lord is mighty, it behooves man to be mightier still. Such a one was Joseph, who was called righteous and guarded in purity the sign of the Holy Covenant which was imprinted upon him.

Rav Elazar said: "The word 'after,' *ahar*,[347] here alludes to the Dark Lord, being the name of the other side. Joseph exposed himself to his onslaught by paying enormous attention to his personal appearance. That gave an opening to the Dark Lord to say, 'Behold! Joseph's father, Jacob, observes mourning for him (Joseph) and he decks himself out and curls his hair!' [Joseph did not, at that moment, feel compassion for his father's anguish over his disappearance and assumed death]. Thus, the bear was let loose, as it were, and set upon him."[348]

The biblical account of Joseph and his brothers has, for most readers of the Bible, been a beautiful and tragic story of family strife and difficulty. The *Zohar*, however, interprets the narrative as part of the cosmic code revealed in the Bible. Astronomy, astrology, sociology, and cosmology are only some of the domains touched upon by the Jacob and Joseph story. Of all the Genesis narratives, those about Joseph are the longest and most detailed. The Joseph account contains an unprecedented wealth of background material that we can use to crack the cosmic code, aiding us in our quest for the Grand Unification Theory.

The Central Column, the cosmic glue for the Grand Unification Theory of modern cosmology, is clearly represented in the Joseph story. The focus of attention in the Genesis narrative is the nobility of Joseph's character and the salvation that came through his compassion. The *Midrash* also softens the harsh treatment that Joseph metes out to his brothers, pointing out that "he had behaved like a brother to them when they were in his power, while they had not treated him like a brother when he was in their power."[349] His brother, Shimon, who had cast Joseph into the pit,[350] was chained up by Joseph, but as soon as the other brothers left, "he gave him to eat and drink, and bathed and anointed him."[351]

Joseph was present at his father's death. At that point, his brothers apparently feared that he would take revenge for their cruel treatment of him in his youth.[352] He dispelled their fears, however, by demonstrating once again his cosmic energy-intelligence of *Yesod* (Foundation) and showing compassion. Joseph had every reason to disregard this attribute of compassion because of how his brothers had treated him, but he acted instead to manifest positive energy. Joseph epitomized the grand unification energy-intelligence, *Rachamim* (Compassion), known as the cosmic *Sefirotic* effect of *Yesod*, which refers to the function of this energy-intelligence within the cosmos. The biblical noun *rachamim* רוֹזְמִים and the verb *racham* רוֹזֵם or *richam*, רְיוֹזֵם which are frequently used to denote this behavior, are derived from the same root as the noun *rechem* רוֹזֵם (womb), leading some scholars of the Bible to suggest that the original meaning of *rachamim* was "brotherhood" because those born from the same womb are brothers or sisters to each other.

For King David, *rachamim* indicated an essential relationship between the Lord (the Lightforce) and those whom the Bible refers to as the Israelites. Those souls originating from the positive energy-intelligence of Abel are known as Israelites, a term that indicates the fundamental characteristic of good. The other peoples of the Earth, whose souls are rooted in the negative energy-intelligence of Abel, are known as *Erev Rav*, the nation of Mixed Multitudes. They are described as unloving and misanthropic individuals.[353] We will discuss these two sides of Abel's energy-intelligence later in this chapter.

"He, the Lord, being full of compassion, forgives iniquity and does not destroy," declares King David.[354] "Learn to do well, seek justice, relieve the oppressed, judge the fatherless, plead for the widow," pleads the prophet Isaiah.[355] These verses have been understood to encapsulate the grand unification energy-intelligence of the Lord, the attribute of compassion. They are meant to establish the norm for human conduct.

"Just as the Lord is referred to as compassionate and gracious, so must you be compassionate and gracious, giving gifts freely.[356] Therefore, you shall keep the commandments of the Lord, to walk in his ways and to fear him."[357]

The *Talmud* expanded and deepened the biblical concept of compassion by recognizing the energy-force of compassion as a vital characteristic of the Israelite, as opposed to the *Erev Rav*.[358] Maimonides declared that arrogant, cruel, unkind people were to be suspected of not being Israelites.[359] In this connection, attention must be directed to the meanings of "Jew" and "non Jew" propounded by Rav Isaac Luria (the Ari). His approach to an ancient misconception actually represents one of the most important constituent elements of Kabbalah. In all the numerous references to this subject in the *Talmud* and the *Midrash*, there is none as startling and revealing as the Lurianic interpretation of this most delicate issue.

But before we explore the Ari's explanation of the origin and nature of the Jew and the non-Jew, we must consider the impact of religion—and religious institutions—on society. The influence of religion on society has been a mixed blessing. There is no denying that many cases of selfless devotion by the religious community have been recorded throughout history. However, many religionists long ago became institutionalized and began to concern themselves more with power and politics than with good and evil. In our own time, religious hatred and hostility fester all over the world. While most religions extol the virtues of "love your neighbor," it is all too often hatred, arrogance, and war that characterize the history of the world's great religious institutions.

Few would deny that religion has set a divisive example in society overall, rather than trying to be a positive force and pulling all peoples toward completion of the world's rectification. The sad history of bigotry became inevitable once religious organizations were institutionalized. As a result, today we see tremendous dissatisfaction with religious establishments in the Western world, and many people have turned elsewhere in their search for spiritual enlightenment and fulfillment.

Kabbalah, through the effort and writings of Rav Ashlag, reveals more of the deeper meaning of existence, of good and evil, than can be found in most traditional religious institutions. Through Rav Ashlag's efforts, we now understand the magnitude of the iniquity committed by those nations and religions that embitter minorities and strip them of their freedom and dignity. Historically, the downfall of nations came about solely as a result of the oppression of minorities and individuals who eventually overcame and destroyed their oppressors. It is now clear to everyone that we cannot establish peace without considering the freedom of the individual.

Thus far, we have defined the individual in terms of how society nurtures him. Yet we must ask: Where is the individual himself? Surely an individual is more than what society gives him and what he inherits from his ancestors. Where is the separate entity we can define as the "self?" The self is central to all Creation. It does not depend upon religion, culture, or politics. The ego-based antagonisms among people, the ever-sharpening international tensions, will not disappear as a result of any political, cultural, or religious stratagem.

In his prophecy of peace, the prophet Isaiah said, "And the wolf shall dwell with the lamb, and the leopard shall lie down with the kid.[360] For the earth shall be full of knowledge of the Lord, as the waters cover the sea."[361] The prophet attributes worldly peace to being full of knowledge of the Lord,[362] and there is the key: knowledge of the Lightforce of the Lord. "All mankind will be united into a single body with a single mind, filled with the knowledge of the Lord,"[363] declares Rav Ashlag.

Let us now turn to Rav Isaac Luria's (the Ari's) new interpretation of the idea of the Israelite and the other nations. Moses said to the Lord, "And if You deal thus with me, kill me, I pray You, out of hand, if I have found favor in Your eyes, and let me not see my wickedness."[364]

Exploring this unusual plea by Moses, the Ari asks why this verse seems to refer to the wickedness of Moses, when, in fact, it was the people who actually appeared to be at fault. The verse, in the Ari's view, should have stated "their wickedness" and not "my wickedness."

The. *Zohar* states that the leaders of the *Erev Rav* (the Mixed Multitudes), were the children of Balaam: Yunus and Yumbrus.[365] The soul of Moses and the soul of Balaam both stemmed from Abel. Although they both issued from the same source, Moses grew out of the energy-intelligence of goodness within Abel. But when the serpent infused Eve with venom (the intelligence of evil), evil and good became commingled in her descendant Abel, and it was from this intelligent energy of evil within Abel that Balaam, the wicked, issued. Because both Moses and Balaam were descended from Abel, Moses concerned himself with converting the Mixed Multitudes from their evil inclination, which stemmed from the evil within Abel. The *Erev Rav* were responsible for the making of the golden calf;[366] they were thus the evil manifestation of Israel, an evil that exists to this very day. The *Erev Rav* emerged from the same source as Moses.[367] Moses and the *Erev Rav* were therefore interrelated.

This, then, was the Lord's advice to Moses: "You go down, for the people [and not the Lord's people] which you brought out of the land of Egypt have corrupted themselves."[368] Moses then pleaded on behalf of the *Erev Rav*, that segment of the nation of Israel that represented the evil of Israel—the evil aspect of Abel, Moses' other half—saying, "Wherefore should the Egyptians speak, and say, for mischief did He bring them out, to slay them in the mountains, and to consume them from the face of the Earth. Turn from Your wrath and display compassion for this evil by the people."[369] The *Erev Rav* were the people of Moses. He brought them into the nation of Israel and pleaded on their behalf for the purpose of cleansing and elevating their evil energy-intelligence so as to make it do only good.

"With this understanding," continues Rav Luria, "an insight will be gained of the coded verse, 'and there never arose a prophet since, in Israel, like Moses, whom the Lord knew face to face,'[370] and of the *Midrash* interpretation that 'within Israel shall never arise one as Moses, but amongst the other nations of the world, there shall appear a prophet as Moses.'"[371] Other nations shall receive prophets like Moses because Balaam and Moses issued from the same source. We can also understand the saying of the sages: "The blessings upon the nation of Israel ought to have been accorded by way of Moses." They were, instead, presented by the leading authority on black magic, Balaam,[372] the purpose being to transmute the energy-intelligence of evil, as portrayed by Balaam—the evil aspect of Israel—and thus to cause a transformation from evil to good.

Two of the Hebrew letters forming the name of *Balaam* בלעם, *Bet* ב and *Lamed* ל, were, by design, the same two letters included within Abel's name הבל. The *Hei* included in Moses' name משה was also the first letter of Abel's name. The Bible indicates the essence of Moses and his origination in the energy-intelligence of good by stating in code: "And the woman conceived, and bore a son [Moses]. And when she saw him that he was a goodly child, she hid him three months."[373] For Moses stemmed from the good aspect of Abel.[374]

This is a most revealing interpretation concerning the nation of Israel: that it consisted of both good and evil. Moses later incorporated this evil within Israel in the form of the *Erev Rav* for the purpose of facilitating the *Gemar HaTikkun*, the Final Correction. The original Noahide Laws[375] were insufficient to achieve the Final Correction. The energy-intelligence of evil, implanted in Eve by the serpent, was far too strong to be overcome by the seven restrictive channels of the Noahide Laws. The task of achieving the Final Correction was a formidable one. Nothing short of the Revelation on Mount Sinai—the giving of the Bible and the precepts—could assure mastery over the Dark Lord and the dominion of good over the energy-intelligence of evil. This is the meaning of the sages' statement that "the commandments were given for the purpose of bringing human beings together."[376]

We now can understand the words of Hillel to the convert. Hillel told the convert that the central tenet of the Bible is: "Love your neighbor as yourself." The rest of the 612 precepts the Lord gave to man are merely preparation and foundation for this single precept. Rav Akiva declared, "This is an important principle in the Bible."[377] The intelligent energy of good becomes physically expressed when men give lovingly to others.

Humankind, as we have shown, is principally divided into two basic categories: the Israelite and the *Erev Rav*. The bloodstained history of religious conflict is a struggle between these two qualities of society: the good and the evil. All religions consist of either those who see "love your neighbor" as the underpinning of their religion or those who have an understanding of the *Erev Rav*, the purveyors of Satan, with his energy-intelligence of evil. Accordingly, when one's own religion can accurately be described as intolerant, bigoted, and vicious, this is a manifestation of the antisocial face of the Dark Lord, the *Erev Rav*. When religious organizations sanction torture, oppression, and genocide, the internal energy-intelligence of evil becomes physically expressed within the cosmos. Religion then becomes a perverting influence, a dark, evil cloud on the horizon of humanity.

Unfortunately, there has been a great deal of misunderstanding regarding the relationship between the Israelites and the rest of the world. One such source of confusion is a prayer, which is contained in the *Siddur*, the Hebrew prayer book: "Blessed are You, Lord, King of the Universe, Who has not made me a *goy* [Gentile]." At first sight, the suggestion seems obvious that to be a *goy* is demeaning and shameful. On closer examination of this most delicate issue, however, evidence to the contrary seems to emerge, especially when we deploy Lurianic Kabbalah. Rav Isaac Luria's radical interpretations enable us to comprehend some of nature's most concealed mysteries. Central to his approach is an understanding of the basic questions of existence.

"And the Lord spoke unto Israel in the visions of the night and said, 'Jacob, Jacob.' And he said, 'Here am I.' And He said, 'I am the Lord, the Lord of your father. Fear not to go down into Egypt; for I will make of you a great *goy* [nation].'"[378] Following on the heels of Israel's sin in making the golden calf, the Lord was determined to put an end to the nation of Israel and begin anew: "And the Lord said to Moses, 'I have seen this people, and behold, it is a stiff-necked people. Now therefore let Me alone, that My wrath wax hot against them, and that I may consume them: and I will make you a great *goy* [nation].'"[379]

The apparent contradiction in these verses is a glaring one, in light of the commonly understood derisive connotations of the word *goy*. And yet it was the Lord's intent to form a *goy* (nation) through Moses. And at the same time, the phrasing of the prayer seems to be yet another contradiction of the deeper philosophical implications of the verse in Exodus. However, before I proceed to reconcile these two apparent contradictory conceptions, permit me to explore another facet of traditional doctrine: the matter of reincarnation.

In *Wheels of a Soul*, I investigated some of the basic concepts of reincarnation, exploring their implications for society and the cosmos. To understand reincarnation, we must first understand that Adam contained all the souls of humanity.[380] His soul became diffused among the whole genus in innumerable codifications and individual variations. All transmigrations of souls are, in the last analysis, only cross-migrations of the one soul of Adam.

For the most part, the kabbalistic worldview transcends the established ideas of psychiatry and psychology rather than merely rejecting them. By observing the universe from another angle, the kabbalist has provided fresh insights and new perspectives on the centrality of man and his place in the cosmos.

In reference to this subject, and for the purposes of our discussion of the *goy*, I recall a little-known section of the Ari's *Gate of Reincarnation*, which describes events beyond the ken of science. The Ari writes: "Laban[381] became incarnated in Balaam, followed by his incarnation in Nabal, the Carmelite.[382] Balaam, the wicked one, drew his power from the serpent and was called upon by Balak to use his power of the tongue (as the serpent) to curse the Israelites. When Balaam was slain, he became incarnated within a stone, the cosmic intelligence of the inanimate, so that his tongue might be silenced. When he ultimately became incarnated in Nabal, the Carmelite, the process of *tikkun* began, for he (Laban) now reached his ultimate destination, the incarnation into an Israelite. When the incident of Nabal's evil tongue towards David occurred, and Nabal was intent on cursing David,[383] he remembered his prior incarnation within a stone for the purpose of correcting the energy-intelligence of evil, and he repented."

In this account, we see that the process of reincarnation is more complex than we might think. To understand clearly just what takes place when a human soul is incarnated, we must review some of the conclusions that Rav Isaac Luria presents on the subject: "The final incarnation, which was Laban's third time around, now crossed over within the framework of Israel. He therefore became the incarnated soul of Nabal, the Carmelite."[384] The Ari spells it out quite clearly: "A non-Israelite may incarnate within the frame of reference of the Mosaic Law. Conversely, an Israelite may incarnate within the framework of the Noahide Laws. Neither framework ensures that the *tikkun* process will take place. A soul in the framework of Mosaic Law finds the *tikkun* process more demanding and consequently more difficult; hence it must obey 613 precepts and not seven." It follows then that for the Ari, the *tikkun* process consists of the individual removing his or her share of *Erev Rav* from the cosmos, whether he or she is an Israelite or not. The extent of the *tikkun* process that one must undergo to complete this task depends on the internal energy-intelligence of one's frame of reference.

Let us now return to where we started. Within the cosmos, there exists the duality of two basic energy-intelligences that originated in Abel: good and evil; the other peoples of our universe and the *Erev Rav*, two nations that are not determined by any particular religion or nationality. These two energy-intelligences become manifest among all nations, religions, and institutions, and hence, the term *goy* (the nation) can represent either of these two characteristics. It is only when good and evil are separated completely that the Messiah will come.

Rav Isaac Luria placed the regeneration of the internal self—which would only come about as a result of obeying the precept "Love your neighbor"—far above the renewal of any religious organization or nation as a political entity. Moral improvement would, in his opinion, bring about the delivery of all peoples from exile, as stated in the verse: "And the Lord shall be King over all the Earth; in that day shall the Lord be One, and His Name One. And men shall dwell in it and there shall be no more utter destruction."[385] The Ari's doctrine of cross-migratory souls imposed the task of *tikkun* upon all peoples of the world.

Regardless of nationality, there are those whose souls originate from within Abel's energy-intelligence of evil. The only effect religion has upon these individuals is to restrict the extent and severity of the evil inherent in their souls. If their souls stem from the cosmic *Erev Rav*, their meanness is without limit. If the *Erev Rav* consciousness incarnates as an Israelite, the potential evil and its effect upon the cosmos could be disastrous, for the religious consciousness of the Israelite is heightened due to the magnitude of his Desire to Receive. The varied genetic codes of humankind depend upon the energy-intelligence of their particular unique aspect of the Desire to Receive. As in the

Endless World, souls are infinite inasmuch as their degree of the Desire to Receive is limitless.[386]

At Mount Sinai at the moment of Revelation, the entire nation of Israel underwent a complete transformation, and the doctrine of *Gilgul Neshamot* (Reincarnation of Souls) became closely associated with the Israelites' role in the universe. It was at Mount Sinai that the Israelites were incarnated with souls of the highest intensity of the Desire to Receive. This was done to enable the Lord to infuse the cosmos with His endless Light and beneficence. The degree to which the Lord would be able to infuse the universe with His power depended upon the capacity, or the desire, of the Vessel to receive this power. To avoid a catastrophic overload of the circuitry of the cosmos, the Revelation of Mosaic Law (the 613 Precepts) was necessary to ensure that humanity could demonstrate and achieve restriction, the cosmic energy-intelligence of the Central Column. Without this restriction, the Israelites' unique quality of the Desire to Receive would simply tap the awesome power of the Lord without channeling or controlling it in any way. In effect, the Israelites might inadvertently become responsible for violence and destruction throughout the cosmos.

The solution to this problem was contained within the Revelation: "Love your neighbor." The Noahide Laws were insufficient to constrain the Israelites' Desire to Receive. Only the Mosaic system of restraint could enable man to fulfill his task in this world if his soul became incarnated as an Israelite. At the time of the Exodus, the souls of the *Erev Rav* displayed an arrogance and insensitivity that caused them to fail to fulfill their task in this world. Such people exist in our age, too: people who lack compassion and sensitivity, corrupting the entire cosmos with negative energy-intelligence and creating a universe torn by violence and destruction—which is no different from the chaos experienced at the time of the Exodus.

Before proceeding further, I would like to explain what I am trying to express when I use this much-abused term: *goy*. When I say *goy*, I intend it to mean what it generally meant before the term became a label for a non-Jew. I mean *goy* to signify insensitivity or a lack of compassion and human dignity, which is the abomination of our universe. In this connection, we must pay close attention to the fact that in all the numerous references to the *goy* in the Bible, the *Talmud*, and the *Midrash*, there is no hint that the term *goy* represents the non-Jew.

The Bible says: "And the congregation lifted up their voice and cried; and the people wept that night. And all the children of Israel murmured against Moses and Aaron. And the whole congregation said unto them, 'Would the Lord that we had died in the land of Egypt or would the Lord we had died in the wilderness. Were it not better for us to return unto Egypt?' And the Lord said, 'I will smite them with the pestilence and disinherit them, and will make you [Moses] a great *goy* and mightier than they.'"[387]

We are obviously dealing here with a situation unique to these people at this time. It is not surprising that the Lord had endured enough of their arrogance. These people were granted freedom, the miracles of the Ten Plagues, the splitting of the Red Sea, the miracle of the manna, and all they could say to the Lord was: "What have you done for us lately?" Such egregious ingratitude prompted the Lord to declare an end to these people of arrogance and insensitivity. Moses would then become the crowning glory of a great new *goy* (nation).

The Israelites' ingratitude was further proof that with the evil energy-intelligence of Abel dwelling within them, they were incapable of completing their *tikkun* in their existing corporeal bodies. Thus the Lord decreed, "Your carcasses shall fall in this wilderness and all that were numbered of you, according to your whole number, from twenty years old and over, who have murmured against me. Doubtless you shall not come into the land, concerning which I swore to make you dwell therein."[388]

"The generation of *Dor De'ah* (the Generation of Knowledge) shall once again rise in reincarnated souls during the Age of Aquarius," declares the Ari.[389] The stain of arrogance and insensitivity shall be expunged from the people, and the doctrine of Hillel, "Love your neighbor," shall one day be achieved. This is the meaning of *tikkun*.

Everything that is done by the individual or the community in the terrestrial realm is reflected in the Upper Celestial Realm. The impulse that originates from good deeds or from the energy-intelligence of Moses, the good aspect of Abel, infuses the entire cosmos with a flow of blessings that spring forth from the Source, the *Mayin Duchrin* of the Lord. Thus, in an effort to restore the original wholeness and remove the fiendish nether worlds of evil, the Lord sent Moses, the aspect of Good, to be the *Goy Gadol*, the Great Nation and Force, to permit the cosmic seed to fulfill its function.

The concern of cosmic *Tet* was directed toward these *Erev Rav* souls in whatever religion or incarnation they existed. She perceived the positive and negative energy-intelligences that ultimately were destined to become physical expressions. Cosmic *Tet* believed that the positive energy of the *Goy Gadol* as represented by the Lord's assurance to Moses, was certain to prevail if she were chosen as the channel for Creation. She considered herself the suitable vehicle to assist those *Erev Rav* souls who would fall to a lower spiritual plane because they were incarnated with the negative energy of the evil of the *goy*. For *Tet* is the first letter and initial energy-intelligence of the Hebrew word *tov*, meaning "good."

A major tenet of kabbalistic doctrine is that the Lord is One, which means that all of life and its emanations are basically good. When the Lord viewed the cosmos as it emerged from the void and darkness, He said, "It is good."[390] Yet this is where we run into one of the basic problems of a monotheistic worldview: How do we explain the existence of evil in its many forms? Extraordinary natural catastrophes and anguish in daily human life are but two of the expressions of evil. To make sense of our worldview, we must explain how these facts fit within the design of Creation.

For the kabbalist, the willful human disregard of the concept of Bread of Shame[391] is the origin and root of evil and corruption. Evil did not originate within the plan of the Lord, but rather, issued from the created being. The created soul's removal of Light brought about the vacuum that allowed for the establishment of darkness and evil.

"The Lord saw the Light and it was good." This sentence parallels the common perception of the energy-intelligence of sunlight, which is generally characterized as giving and sharing. The framework of the Lord is sharing and compassion. This stands contrary to the energy-intelligence of evil, which is directed by the Desire to Receive for Oneself Alone.

The numerical value of cosmic *Tet* contributed to her plea to be the cosmic channel for Creation. To understand this matter, we must rely on *gematria*, or numerology, one of the kabbalistic methods for interpreting the Bible and revealing its cosmic code.

Let us therefore employ the *gematria* in deciphering the internal energy-intelligence of cosmic *Tet*. The letter *Tet* ט embodies the ninth letter of cosmic *Binah*—*Yesod* of *Binah*— or number nine. Consequently, cosmic *Tzadi* צ, representing the ninth decimal letter of cosmic *Zeir Anpin*—the number ninety relative to cosmic *Tet*—is the externalization of *Tet*. Cosmic *Tet* represents the internal, concealed force of cosmic *Tzadi*, which is *Yesod* of *Zeir Anpin*, the body and soul. The Hebrew word *goy* גוי encompasses the energy-intelligence of most of mankind. The number nine indicates the affinity of those words with cosmic *Tet*. Consequently, the internal energy-intelligence of cosmic *Tet* determines the characteristics of practically all humanity.

This is what cosmic *Tet* had in mind when she made her plea to the Lord. "The interest of the entire universe would best be served if I served as the cosmic channel for Creation," cosmic *Tet* said on her behalf. "The Dark Lord would never undertake any attempt to engage the Lord's Lightforce, for he could never survive an onslaught by my internal energy-intelligence."

Cosmic *Tzadi*, the station for cosmic *Tet*, is called "Righteous," for she connects with and augments the forces of *Malchut*, the

World of Action. Cosmic *Tzadi*, with *Tet*'s energy-intelligence aboard, was much stronger than the evil, arrogance, and insensitivity that formed the weaponry of the Satan. With the combined forces of cosmic *Tet* and cosmic *Tzadi* as protection, humankind could not fail in battle with the Dark Lord. Any attempt by the Satan to restore his rulership over the cosmos was destined to fail. Through the power of *Tet*, the entire cosmos could become paradise in one clean sweep.

Tet, cosmic *Yesod* of *Binah*, constitutes the internal intelligence of *Tzadi*, *Yesod* of *Zeir Anpin*, and bears the attribute of *tov* (good), indicated by its portal opening at the top of the spacecraft. Cosmic *Tet*, in fact, is the only letter of the *Alef Bet* with a direct connection from its upper structure to the Lord.

The prophet Isaiah hints at the existence of this mysterious spacecraft when he exclaims, "Say to the righteous, that it shall be good with him."[392] Furthermore, the mystery of cosmic *Tet*'s awesome power—her concealed Light—is the Light the Lord created on Day One of Creation. The sages of the *Talmud*, alluding to the cosmic power of *Tet*, declared that "this Light enabled Adam to see from one end of the world to the other."[393] This bright flame of power would ultimately cast the light of new truth across the galaxies.

"And the Lord saw the light, that it was good."[394] One who sees the letter *Tet* in a dream is indeed blessed. Why is this so? The Bible used the letter *Tet* in the words: "that it was good." Thus, *Tet* is understood to possess the Light that radiated throughout the universe from one end to the other.[395] This is why cosmic *Tet* believed herself to be the proper Vessel for the creation of the world.

The Lord, however, exists beyond the curtain of time and foresees the future. Upon seeing the wicked generations of the Deluge and the Tower of Babel, the Lord had doubts regarding cosmic *Tet*'s intrinsic power of the Light. Thus, the Lord decided to hide his Light for the future. The Light would be revealed for the *tzadikim* (righteous people), who will appear when the Final Redemption is consummated. This event is revealed in the words: "and the Lord saw the light, that it was good."[396]

Cosmic *Tet* possessed the energy that permitted Adam to see from beginning to end—truly an awesome power of intelligence. She was immune to the evil force of the Dark Lord. However, the *Erev Rav*, the *rasha'im* (wicked), could crack her security shield by means of their corrupt deeds, thereby placing the evil *klipot* in a position to seize the Lightforce.

So long as cosmic *Tet* remained the property of the righteous, they were in a position to make full use of this power, as Adam had done before them. And certainly, if she were chosen to be the channel for Creation, the righteous could count on cosmic *Tet* as part of their arsenal. However, in a universe of free will, the *Erev Rav* could corrupt the deeds of the righteous, and indeed, the Bible is full of accounts of the *Erev Rav*'s ugly acts. The *Erev Rav* would thus inadvertently provide the *klipot* with the opportunity to seize the power of cosmic *Tet*. This potential for a breach in her security shield caused the Lord to reject *Tet* as a suitable channel for Creation. Instead, her mere presence within the cosmos has become a symbol of mourning and sadness.

The letter *Tet*, representing the number nine, has become a symbol for worldwide crisis. The *Talmud* decrees *Tisha B'Av*

(Ninth of *Av*) as the major day of mourning because calamities occurred on this day again and again throughout history.[397] Unfortunately, the violence and disruption that affected the people have brought pain and sorrow to all nations of the world. On the Ninth of *Av*, it was decreed that the children of Israel, those liberated from Egypt, should never enter the land of Israel.[398] The First Temple, built by King Solomon, was destroyed by the Babylonian king Nebuchadnezzar on the Ninth of *Av*. "And all the congregation lifted up their voice and cried. And the people wept that night.[399] When the sound of their weeping reached Heaven, the Lord said, 'You weep not without cause. The time will come when you shall have good cause to weep on this day.' It then was decreed that the [Second] Temple be destroyed on this same day, so it became forever a day of tears."[400]

The Second Temple was destroyed by the Romans some 670 years later. Bethar, the last stronghold of the leaders of the Bar Kochba war, was captured on this date. The expulsion of the Jews from Spain in 1492 is also said to have occurred on the Ninth of *Av*. On the Ninth of *Av*, the Israelites even refused to enter the land of Israel because they feared the powerful nations that controlled Israel, a fear communicated to them by the spies they sent to explore the territory.

The biblical book the rabbis designated as the cosmic connection to this unique day, *Tisha B'Av*, is *Eicha* אֵיכָה, or the Book of Lamentations. The word *eicha* is similar to the word *eiyeka* אַיֶּכָּה (where are you), which the Lord cried out to Adam in grief and anger after the sin in the Garden of Eden. Both words are composed of the same consonants; only the vowels are different. These words are aspects of the biblical cosmic code, as are all the words of the Bible. Both words are connected with exile, and they represent disaster and destruction. Just as the sin of Adam culminated in his removal from the center of energy and immortality when he was expelled from the Garden of Eden, so was Jerusalem destroyed for the sins of Israel, and the Israelites were banished from the land of Israel and sold into captivity and slavery.

As strange as it may seem, this day also will see the birth of the Messiah—the beginning of the Final Redemption. In this connection, we must search for the deeper significance of these catastrophic events that have shaken our universe. Why did these events happen on the ninth day and not another day? What is the cosmic force behind this seemingly difficult time period? Why have we been unable to deal effectively with this problem?

The connection between cosmic *Tet* and this peculiar ninth day is quite apparent. It is a Universal Law that for every power, there exists an equal and opposite power. Otherwise, there could be no free will in the universe. If cosmic *Tet*'s plea to the Lord was based on her energy-intelligence of goodness, then an equal force of negativity must have also existed. King Solomon, concerned with the Universal Laws that guide our lives, clearly defines for us the essential truth surrounding the essence of Creation: "In the day of *tov* [good], be connected with *tov*. And in the day of *ra* [evil], consider the Lord has set the one against the other."[401]

In his interpretation of the cosmic order, King Solomon provided the fundamental cosmic code concerning *Tet*. The deciphering was completed by Rav Shimon bar Yohai in the *Zohar* (*Book of Splendor*). Rav Shimon's interest in astronomy points to his fundamental awareness of the celestial dimension of mankind. Concerning one of the most difficult and abstruse sections of the Bible, Rav Shimon revealed the essential consciousness and energy-intelligence of the ninth day of *Av*: "And Jacob was left alone; and there wrestled a man with him until the breaking of the day. And when he saw that he prevailed not against him, he touched the hollow of his thigh; and the hollow of Jacob's thigh was strained, as he wrestled with him. Therefore, the children of Israel eat not the sinew of the thigh-vein, which is upon the hollow of the thigh, unto this day, because he touched the hollow of Jacob's thigh, even in the sinew of the thigh-vein."[402]

The sciatic nerve, together with other arteries and tendons, must be removed from the slaughtered animal before that portion of the animal can be ritually prepared for consumption. This is a constant reminder of the Divine providence extended to Jacob. Although there is no doubt regarding the content of this precept, I would still like to ask whether we are to understand it simply as a historical phenomenon, or whether we might consider Jacob's experience to be a coded message relevant to our own existence. The kabbalist sees the Bible as containing coded information waiting to be deciphered. The Bible comprises much more than a simple literal translation. Kabbalah reveals a wealth of concrete historical facts, as well as an understanding of the Bible's deeper significance. It is connected with, and inseparable from, a certain stage of religious consciousness. A kabbalistic interpretation lies at the root of every biblical story and precept.

A good starting point for our investigation can be obtained by scrutinizing the *Zohar*, which places emphasis on a direct and intimate consciousness of the celestial, metaphysical realm. This is religion in its most acute, intense, experiential state.

"And Rebecca, his wife, conceived. The children struggled together within her; and she said, 'If it be so, wherefore do I live?' And she went to inquire of the Lord. And the Lord said unto her, 'Two *goyim* are in thy womb. And two peoples shall be separated from your bowels. And the one nation shall be stronger than the other. And the elder shall serve the younger.'"[403]

How are we to interpret this dialogue? This is the great enigma the kabbalists and the commentators on the Bible have tried to solve. For it must be said that this direct relationship between Rebecca and the Lord—between the finite and the infinite—is highly paradoxical. How can words express an experience for which there is no adequate parallel in this finite world of man?

It would be superficial and incorrect to conclude that an enigma implies an absurdity. It would be wiser to assume, as the kabbalistic view does, that the religious world of the mystic can be expressed in terms amenable to rational knowledge. Therefore, we must pay attention to the *Zoharic* perspective by which we recognize the basic cosmic intelligences, good and evil, pervading the universe, which we see in the *Zohar* as follows:

"And the children struggled within her, for in her womb already Esau declared war against Jacob. Observe that the one was the side of him who rides the serpent, while the other was of the side of the Lord."[404] These two *goyim* were to represent an ongoing struggle between good and evil—peace and tranquility versus turmoil and destruction. The sinister demonic world of evil is nourished and quickened by the sin of man. Evil is intelligence making illegitimate inroads upon the Celestial Realm of Light.

This drama of history brings us back to the problem of the sciatic nerve as it relates to the Ninth of *Av*, the most complex and confusing day of the year. What is the basis for this precept? How is this precept to be regarded within the framework of the biblical cosmic code? For an in-depth understanding of this concept, let us return to the *Zohar*.

Why is the nerve in question called *gid ha'nasheh* (sciatic nerve)? The sciatic nerve contains the power and intelligence to swerve man from the Lord. In that nerve resides the intelligent energy of evil. And because the extraterrestrial intelligence that fought with Jacob could not find any other weak spot in Jacob's body, it connected with its primary battle station within man: the sciatic nerve. Consequently, the Bible forbade the eating of this nerve. The body of man connects with the metaphysical realm. If a body part is good, it draws the energy-intelligence of good, but an evil body part draws the energy-intelligence of evil.

Similarly, when a man eats the meat of an animal, the intelligent energy of the body part of the animal connects with and is metaphysically integrated with the corresponding body part of the man. For this reason, Israel does not eat the sciatic nerve so that the internal energy-intelligence of the nerve will not be injected into the corresponding nerve of man. This is one way to prevent the energy-intelligence of evil from becoming physically expressed within the cosmos.

There are many connections between the body of man and the metaphysical principles undergirding the cosmos. For instance, there are 248 positive precepts in the Bible, corresponding to the 248 bones in the human body. This number of precepts was purposely designed so that each precept might connect directly with the body of man, thus providing the individual a connection to the energy-intelligence of the Lord. Similarly, there are 365 prohibitive precepts of the Bible, corresponding to the 365 nerves (including the sciatic nerve) and the 365 days of the year. *Tisha B'Av* (the ninth day of *Av*), is the day corresponding to the energy-intelligence of evil that is related to the particular intelligence of the sciatic nerve.

It is written: "The children of Israel eat not of the sciatic nerve." In the Bible verse stating this commandment, the seemingly nonessential word *et*[405] אֶת is inserted. The two letters composing this word, *Alef* and *Tav*, reveal the reason for fasting on this particular day. Just as the injunction against eating the sciatic nerve prevents the energy-intelligence of evil from becoming physically expressed, so, too, is it necessary to fast on the Ninth of *Av* because the energy-intelligence of evil pervades the universe on this particular day.

The Bible, the cosmic code of the Lord, reveals these principles through the tale of the patriarch Jacob wrestling with the angel, who represents the energy-intelligence of evil. "And there wrestled a man with him"[406] refers to the perennial struggle of humankind with evil. The angel could find no place to connect with Jacob other than his (humankind's) sciatic nerve. This is the meaning of the verse: "And when he saw that he prevailed not against him, he touched the hollow of Jacob's thigh and his thigh was strained."[407] At that moment, Jacob became weak. Similarly, the evil intelligence found that the ninth day of *Av* was a day when it could prevail. The result was the destruction of the Holy Temples because of man's corrupt deeds and weakness. The destruction of the Temples also caused a disruption in world peace and tranquility.

Those who eat on *Tisha B'Av* are likened to those who eat the sciatic nerve. Rav Chiya said in the *Zohar*: "Had Jacob's strength not failed him at that spot (the sciatic nerve), he would have prevailed against the angel so completely that Esau's power would then have been broken, both on the terrestrial and extraterrestrial realm."[408] The *Zohar*'s explanation of the Bible is crucial to a more profound understanding of the cosmos. The Bible must be conceived as a vast *corpus symbolicum* of the entire world. Out of this cosmic code, the inexpressible mystery of the Celestial Realm can be deciphered. The dogmatic rituals and precepts commanded by the Bible are, to the kabbalist, symbols of a deeper, concealed reality. The infinite becomes revealed through the finite. The soul radiates throughout the body, infusing the body with life. Without the Lightforce, reality ceases to exist.

This brief summary provides us with some idea of the fundamental difference between interpreting the Bible superficially and understanding it as a cosmic code. If the whole of the universe is to be considered an enormous, complex machine, then humankind is the technician that keeps the wheels turning by providing fuel via the energy-intelligence of its actions. The presence of humankind is thus of central importance. Every precept becomes an event of cosmic importance. The energy-intelligence of each precept has a bearing upon the dynamic interplay of the universe. When every precept is interpreted as a cosmic event, a seemingly insignificant precept such as the prohibition against the eating of the sciatic nerve, assumes a much greater significance.

Seen in this light, there is nothing more instructive than our comprehension of cosmic *Tet*. To begin with, *Tet* embodies the power that permitted Adam to see from beginning to end. *Tet* encompasses the wide range of human behavior, including all the *goyim*. Her numerical value of nine finds expression in the cosmic effect of *Tisha B'Av*, a day of destruction and mourning. The depth of her penetration into the hidden worlds of the cosmos can be encountered at every level. So there was perhaps no letter more suitable than cosmic *Tet* as the channel for Creation, in that she represents the *tov* (good) that exists in the cosmos.

The Lord, however, had another fate in mind for *Tet* when He replied, "I will not create the world through you, *Tet*, because the goodness you represent is concealed within you." This truth is revealed by the psalmist when he declares, "O how abundant is Your goodness which You have laid up for those that fear You."[409] And so the Lord continued "This hidden goodness has no place in the world I am about to create, but only in the World to Come."[410]

Paradox is and always has been an essential distinguishing characteristic of the Kabbalah, and the understanding of cosmic *Tet* is no exception. She is characterized by two contradictory features that are nevertheless interrelated. The first of these is her immense power and its energy-intelligence of *tov* (good). Yet her distinguishing trait of goodness required a correspondingly extreme defensive mechanism. Hence, there was a need to conceal her from the wicked, which is the second of her contradictory traits: *Tet* is both immensely powerful and deeply hidden.

"Your goodness is hidden within you," the Lord said, "which is why your goodness is not destined for this world, but rather for the World to Come."[411] Only *Tet*'s externally apparent, diminished energy-intelligence—and not the intense energy-intelligence of her essential intrinsic nature—was ordained for inclusion within the cosmic process. The limitation of cosmic *Tet* mirrors the limitations placed upon humankind. The initial good of Creation was to unseal the soul and untie the knots that bound it. All internal forces and concealed soul-energies in humanity are still distributed and differentiated in the corporeal energy-intelligence of our bodies, but one day, duality and multiplicity shall disappear in the all-embracing unified Lightforce of the Lord. The *Gemar HaTikkun* will then be the final liberation of the souls.

For the present, however, certain barriers exist that separate the essential power characteristic of cosmic *Tet* from the stream of cosmic life, a security shield that keeps *Tet* confined within the normal borders of the Lord's battle station and protects her against the flood of negativity released by the *Erev Rav*, the wicked. The Lord has impressed "seals" upon cosmic *Tet* that protect her against any negative energy-intelligence and guarantee her normal functioning. Of special interest in this connection is the doctrine of Adam's sin and his subsequent banishment from the Garden of Eden and immortality.[412] What do these facts represent when interpreted kabbalistically? What is the coded message contained within the phrase "Fall of Adam?" Why were Adam and Eve suddenly conscious of their nakedness, and why did they sew fig leaves together?[413]

In his original paradisiacal state, Adam had a direct relationship with the Lord within the consciousness of *Zeir Anpin*. Adam had what we have been describing as an outer-space connection. Man's creation involved a synthesis of all the spiritual forces that participated in the creation of the universe. The internal consciousness of the Lord's omnipotence is reflected in Adam's organism. Originally, Adam was a purely spiritual energy-intelligence. His ethereal form was patterned after the pure, naked, raw, energy-intelligent Thought of the Lord. Adam's state of consciousness existed within the cosmic realm of *Zeir Anpin*. At this level of consciousness, Adam was immortal.

Before the fall of Adam, Heaven and Earth were of one thought and in perfect harmony. The wellsprings and channels through which energies from the Higher Celestial region Region flow into the Lower Realms were still completely active. The Vessel (Adam) and the Lightforce were still in perfect tune with each other.

But when Adam sinned, he descended into the realm of corporeal existence. He was born into the pollution of all physical matter, and mankind's outer-space connection was thereby severed. The order of things was turned into chaos because the Lightforce was simply too hot to be handled by a physical being. Raw, naked energy of this intensity was not meant for our World of Action. The Thought-process exists beyond the limits of time and space. The corporeal realm, with all of its limiting factors, was unable to channel the Heavenly communication. Consequently, the biblical code continues: "They sewed fig leaves together and made for themselves *hagorot* [insulated garments]," which would allow them to withstand the primal energy-intelligence of the Lightforce, just as astronauts need spacesuits to protect themselves from the perils of outer space.

It is here that we discover why cosmic *Tet* was not suitable as the channel for Creation. The cosmic energy of her internal structure was so intense that a polluted, mundane process was incompatible with her outer-space-oriented intelligent energy. The frailty of the terrestrial realm required that *Tet*'s innermost recesses be concealed to prevent chaos from being unleashed upon the universe.

The concept of nakedness now becomes closely aligned with the concept of the outer-space connection. The individual soul retains its own particular existence only to the extent that it can sustain the energy-intelligence of the Lightforce. When Adam and Eve were cosmically severed from the Thought-process of cosmic *Zeir Anpin*, they were no longer in accord with the place they had originally occupied; they were no longer part of the dynamic interplay of the whole. Their inability to handle the intensity of the Lightforce left them naked. Their banishment into the exile of a strange new form of corporeal existence was a particularly abhorrent punishment, both spiritually and physically. Fig leaves, which provided insulation from the unadulterated energy of the Light, were their only protection.

The remedial properties of herbs have been recognized and appreciated since time immemorial. Lately, the fast-acting chemical solutions provided by science have diverted our attention from herbal remedies and the nature of true healing, but the use of herbs still continues among certain cultures and communities. The use of herbs is the oldest medical science.

Herbs are mentioned in the Bible from the beginning of Creation. "Thorns also and thistles shall it bring forth to you; and you shall eat the herb of the field," the Bible says.[414] When Adam and Eve were driven from the Garden of Eden and had no more access to the Tree of Life, they were vulnerable to the polluted terrestrial realm of corporeality. Therefore, the Lord added herbs to humankind's diet to help protect it against terrestrial afflictions. As the prophet Ezekiel declared, "The fruit thereof shall be for meat, and the leaf thereof for medicine."[415] These are the Lord's remedies, necessary for mundane existence.

Originally, everything in the universe was expressed as a unified, intelligent energy whole.[416] The universe was one indivisible, dynamic entity whose intrinsically interconnected parts could be comprehended only as a pattern of Thought in the grand cosmic process. But sin created a severance from the Lord's omnipresence, causing the loss of the original unified whole and leading to an isolated, fragmented existence. In the realm of the outer-space connection, one never defines entities or concepts as "things." One deals always with the interconnectedness of the Thought-process. Within cosmic *Zeir Anpin*, nature appears as a complicated web of thought-relationships between the various parts of an all-embracing unified whole. But for the intervention of evil, the universe would never have assumed a fragmented, material form.

The threat of nuclear war is the greatest danger facing humanity today. This awesome power is obtained by breaking uranium nuclei into fragments. Man has learned how to shatter the very foundation of unity within our universe. What are the components of an atom, if not thought? The creation of a fragmented atom has led to a profound environmental imbalance, generating numerous symptoms of ill health and ill will. The division of the atom is precisely the sort of disrupting scenario created by the sin of Adam. He, too, created a division between the realm of cosmic *Zeir Anpin* and the energy-intelligence of this world, *Malchut*.

If the internal energy-intelligence of cosmic *Tet* were seized by the wicked, the result would be a holocaust far more disastrous than any nuclear war. Therefore, cosmic *Tet* could not become the channel for Creation. The time for her intelligent energy to emerge was destined to be at the *Gemar HaTikkun*, the Final Correction. Cosmic *Tet*'s dimension in her concealed state was insufficient to correct this world, cosmic *Zeir Anpin*, and *Nukvah* (*Malchut*). As a result of this inadequacy, cosmic *Tet* would be an easy mark for an onslaught by Satan.[417]

The Lord continued speaking to *Tet*. "Furthermore, your concealment was the reason for the Gates of the Holy Temple disappearing into the ground."[418] The gates disappeared into the ground so that they would not be seized by the Romans (the physical expression of the Dark Lord's fleet).[419]

This is where cosmic *Chet* 𝁉 entered the scenario as a counterbalance to cosmic *Tet* ט. Why and how did *Chet* play this role? As *Resh* and *Kuf* did before her, *Chet* served a dual purpose within the cosmos. *Chet* represents the energy-intelligence of cosmic *Hod*, which corresponds to the Left Channel bequeathed to the cosmos by *Yesod*. Cosmic *Tet* serves as the Right Channel, a cosmic embryo by which the souls are brought to birth. The Left Channel—cosmic *Chet*—excretes waste to the *klipot*. Cosmic *Chet* is identified with cosmic *Kuf* because the latter resides, like *Chet*, within *Yesod*. *Yesod* is the celestial channel that provides the entire cosmos with the various intelligent energies necessary for its existence. As mentioned previously, the portion of energy reaching the Dark forces is furnished by the *Kuf*. Hence, *Chet*, like *Kuf*, appears sometimes as an entity of the Lord's twenty-two intelligent energies, while at other times, she is in the service of the Dark Lord. This dual nature of *Chet* takes the form of good and evil.[420] While cosmic *Chet* served the cosmic process in Creation,[421] she also undertook "to excrete the waste to the forces of Darkness." These two channels, *Chet* and *Tet*, are only separated by a very thin wall, similar to the papery skin that clothes garlic.

In a like fashion, the male sex organ consists of these two intelligent energy-forces: the life-intelligent force of the sperm and the excretion of the urine. Cosmic *Tet* is the intelligent energy channel that provides a bridge for souls traveling from prior lifetimes into the present. We have already seen *Tet*'s unique ability to unite the past and the future by providing Adam with an all-pervasive view of the universe from the beginning to the end. Transpermia, the term coined by Francis Crick, the famous discoverer of DNA, is actually cosmic *Tet*'s intelligent energy-force, which provides the physical expression of an incarnated soul along with its DNA. *Tet* joins the past with the future.

Cosmic *Chet* serves the Satan and the *klipot*. It is therefore no coincidence that when one speaks the letter *Chet*, the letter *Tet* combines with *Chet* in its pronunciation. Additionally, the word *chet* חטא means "sin," which consists of the two letters *Chet* and *Tet*.

The Left Channel is sometimes capable of gaining dominance over the Right Channel. For precisely this reason, the male organ also has been employed for evil purposes, such as rape. The two letters combined as one unit in the word *chet* (sin) comes as the result of this dominion. Moreover, the numerical value of the word *chet* (cosmic *Chet* plus cosmic *Tet*) is seventeen, which is the same value as *tov* (good), implying that good and evil stand one against the other. King Solomon referred to this phenomenon when he declared, "The Lord has set the one against the other."[422] So when the Right Channel, where cosmic *Tet* is lodged, gains dominance over *Chet*, the scale is tilted in favor of *tov* (good). On the other hand, when the Left Channel of cosmic *Chet* dominates over *Tet*, the energy-intelligence of *tov* is subdued by *chet* (sin).

This, too, is why the Lord refused cosmic *Tet*'s request that she serve as the channel for the creative process. Her presence would give the *klipot* of the Dark Lord the power and opportunity to tap the Lightforce of the Lord for themselves. Thus, if man corrupted his way, the power of the cosmic *Tet* would come under the dominion of the Dark Lord for eternity.

In this connection, let us examine two of the two coded words under consideration: *tov* and *chet*. Deciphered, they incorporate the entire dialogue between the Lord and the two respective cosmic channels, *Tet* and *Chet*, concerning the cosmic process of Creation. The word *tov* consists of three letters: *Tet*, *Vav*, and *Bet* טוב . Numerically, they add up to seventeen. Inasmuch as *Tet* is the head of this channel of letter-energy, she is therefore in a position of dominant, primary power. The secondary and tertiary letters of the energy channel of *tov* are *Vav* and *Bet*, which have the numerical value of eight—the same as *chet*, sin. Consequently, when cosmic *Tet* (numerical value of nine) gains dominance over *Chet* (numerical value of eight), the result is *tov*, good, for the reason that nine is greater than eight.

Conversely, cosmic *Chet*, in her coded position within the word *chet*, indicates her dominant position over cosmic *Tet*. The consequence of the defeat of *Tet* is sin. Because *Chet* is the first letter of this word-channel of energy, *Chet*'s position in this word is dominant. The secondary letter of *chet* (sin) is the letter *Tet*. Eight precedes nine, which means that in *chet* precedes goodness. The overall result is *chet* or sin, which is contrary to *tov* or good.

If there is hatred, violence, and disorder among human beings, it is because these elements exist within the fragmented body of humankind. We give the Dark Lord the right of entry through our selfish Desire to Receive for Oneself Alone. Why do bad things happen to us? Because we have become aligned with cosmic *Chet* and our actions are counterproductive to *Tet* or *tov*. Our Desire to Receive for Oneself Alone makes us our own worst adversaries.

Most of us are trapped in our self-centered cosmic egg, which severely limits our human potential. It is difficult to accept the notion that when we want to receive, we must share.[423] To receive *tov*, one must share and give to others. This paradox shapes our perceptions of all things around us. When sin or *chet* is the primary process for our energy-intelligence, then we exclude *Tet* or *tov* from our cosmic system. Bad things happen to us because we make them happen. To perceive *chet* (sin) as a reality instead of an illusion is to dwell in a contradiction.

For this reason, the *Zohar* says, the letters *Chet* and *Tet* are not found in the names of the twelve sons of Jacob.[424] By this omission, the biblical code indicated that the tribes stem from a higher and more concealed level of consciousness, thus separating them from cosmic *Chet*, who is a primary source of negativity. This is why the sages of the *Talmud* said, "The bed of patriarch Jacob is perfect,"[425] meaning that energy-intelligence of a negative nature never issued from Jacob to the Dark Lord. This was not so for the other two patriarchs, Abraham and Isaac.

Cosmic *Tet*, upon hearing the Lord's reply, understood her duty, painful though it was to her. She realized the opportunity the Dark Lord would have if humankind were to corrupt her channels in the cosmic process of Creation. Having no other choice, she departed.

Cosmic *Chet* made no attempt to convince the Lord of her unique ability to act as a channel for the creative process. The Lord's reply to cosmic *Tet* clearly stipulated the danger of *Chet*'s presence in the cosmos. Her intelligent energy was available when and if the Dark Lord found it necessary to make a cosmic connection, as was the case with cosmic *Resh* and *Kuf* before her. Hence, cosmic *Chet*'s involvement in Creation was also out of the question.

CHAPTER 18

THE LETTER
ZAYIN

LET MAN ALWAYS CONSIDER
HIMSELF AS THOUGH
HE WERE HALF RIGHTEOUS
AND HALF WICKED.

—*TALMUD, KIDDUSHIN*

Cosmic *Zayin* weighed her options. *Tet*, despite her power, had been rejected. Were the Dark Lord to have seized *Tet*'s power by reason of man's avarice, no chance would remain for humanity to achieve the *Gemar HaTikkun*. The Lord could not afford to gamble on the free will of humankind. *Tet*'s moment of glory would have to wait until the Final Correction.

The odds seemed stacked against the positive aspects of future existence. The Dark Lord knew all too well the impulses that would compel man's greed and corruption. So long as the Desire to Receive remained unfulfilled, greed would continue to be an inherent characteristic of mankind. In fact, the Desire to Receive had to be, of necessity, a permanent fixture in the World to Come because if humanity's challenge of ridding itself of that evil inclination were to suddenly disappear, so, too, would the purpose of Creation, which was to give mankind free will sufficient to remove Bread of Shame.

Cosmic *Zayin* foresaw man's mad race for acquisition and she knew the tenacity with which he would approach the problem of corporeal attainment. She saw the evil that he would attempt to justify in the name of survival. She understood that the vast majority of the Earth's inhabitants in the World to Come would try to satisfy the Desire to Receive for Oneself Alone with material wealth and creature comforts. She perceived that humankind would lack the power to rid itself of this evil inclination. She saw all this, and it troubled her deeply. But what if some cosmic intelligent energy were to be found that might reduce or even neutralize the energy-force of the Desire to Receive for Oneself Alone?

With this view of a future cosmos, cosmic *Zayin* approached the Lord to enter her plea to be the celestial channel for Creation. An energy-intelligence that would later manifest as the Holiness of *Shabbat* accompanied *Zayin*, for she would later be the channel through which the *Shabbat* would come into being.

The universe quaked, and light swirled around the Throne. Cosmic *Zayin* bowed humbly before the Lord and said, "If it please You, O Lord, create the world through me. It is my energy-intelligence by which the *Shabbat* will manifest. And will it not be written: 'Remember the Sabbath day to keep it holy'?[426] Choose me, O Lord, for then the universe will be assured of peace and tranquility."[427]

From the kabbalistic perspective, each of us is an integral part of the grand cosmic design. Whenever we experience ourselves as separate from the rest of the cosmos, we are falling victim to a dangerous illusion, an addictive myth that can cause us to pursue personal gain at the expense of others and to crave more and more intensely to satisfy the Desire to Receive for Oneself Alone. This error can cause us to narrow our focus to include only petty personal concerns. Living this way can alienate those closest to us and distort our perceptions, not only of ourselves and others, but even of time and space.

The kabbalist seeks to transcend this addictive inclination by extending his or her consciousness and expanding the range of his or her vision to include others. Only then is it possible to experience oneself as part of the infinite design. By enlarging our circle of sharing to embrace all that lives— which, according to the kabbalist, includes everything in the universe, both animate and inanimate—we begin to rise above physical illusion and experience the boundless beauty of cosmic unity.

Kabbalistically speaking, "desire" is not an entirely negative trait in and of itself. The curse of mankind is not desire itself, but only the negative aspect of desire: the Desire to Receive for Oneself Alone. Desire in the sense of striving toward some positive purpose is, in fact, the very cornerstone of spiritual, personal, and material liberation, as well as the key to inner peace and security. Even the saintliest of men and women are filled with desire—more desire than could be imagined by those whose cravings encompass only greed and acquisition.

Desire, however, becomes a virtue only when it is tempered with an underlying motivation of sharing. Unfortunately, any review of history makes it immediately and painfully evident that the vast majority of people are not motivated by a desire to share. Very few are able to escape from the negative aspect of desire, because to do so, one must possess a consciousness of sharing, an attribute that is excluded or contradicted by humankind's inherent Desire to Receive for Oneself Alone. While a few labor at the monumental task of elevating human consciousness, many more work to lower it, the result being that each step forward on the grand scale of human evolution is accompanied by two steps back. While it is true that science, the arts, religion, philosophy, capitalism, socialism, communism, ethical humanism, and a host of other "isms" may each in its own way have contributed to the awareness of certain individuals, and while these ideologies may have even improved the living conditions of segments of the general population, humanity as a whole seems no closer today to achieving what is, or should be, our true objectives: harmony, freedom from want and hunger, and world peace.

Pitted against the Dark Lord, humankind has not fared well. Despite humanity's concerted efforts to achieve peaceful coexistence, Satan continues to maintain discord within the universe. No plan for peace seems adequate to lessen the tension between the nations of the world. War, strife, and pestilence still reign supreme. Human history stands as an indelible reminder of the failure of good ideas to penetrate the human heart. The precepts of the major religions, for example, have been articles of faith for centuries, yet the benefits of religion appear minuscule in the shadow of the horrors of the wars fought in its name. The ongoing holocausts of genocide, global war, mass murder, torture, and terrorism—not to mention the atrocities committed by humankind against the Earth and her precious resources—forewarn of a troubling future, as negative forces continue to emerge from the depths of the human psyche.

The persistence of such difficulties constitutes an overwhelming indictment of all our cherished beliefs and institutions. New creeds and doctrines arise to address these issues, but no sooner are they expressed than some segment of humanity again commits some new outrage, leaving the rest of us to watch in horror as the whole ideological edifice collapses like a house of cards. Then, slowly, the quest for redemption begins anew. The excruciating irony of it all is that even with hard evidence staring us in the face, we still persist in maintaining destructive illusions. Ideas alone will never transform human behavior. Only consciousness is capable of penetrating to the essence of the human condition.

In light of this perception of the frailties of humankind, cosmic *Zayin* foresaw that neither religion nor the advanced intelligence of science would lead man to his *tikkun*, a term that signifies the completion of the spiritual process of correction and thus has specific meaning for the incarcerated soul. Cosmic *Zayin* foresaw chaos and violence in the life of humankind on Earth, and she contemplated the senselessness of it all. Humanity could be surrounded by laws, edicts, commandments, and moral injunctions, yet these would mean next to nothing unless the negative aspect of desire was prevented from ruling the human psyche. Unless that one seemingly intractable problem could be overcome, the emotions that would drive humankind would forever direct the cosmos in totally negative ways.

Cosmic *Zayin* recognized that violence and disorder were the Dark Lord's objectives and that he was indeed a formidable adversary. The solution, as she saw it, was to impregnate the cosmos with a positive energy-intelligence sufficient to reduce the unlimited craving of the selfish aspect of the Desire to Receive. No greater goal could be found than to establish a universal system of sharing that would cause humankind to contemplate its individual responsibility to and within the cosmic unity.

When the force of *menucha* (tranquility) can begin to articulate itself as a positive influence in the world, it will be largely through the intelligent energy of cosmic *Zayin*. She alone symbolizes the channel by which Israel, in observance of the *Shabbat*, pervades the universe with *menucha*. The intelligent energy of cosmic *Zayin*, portraying the concept of tranquility, is coded in the verse: "Zachor [remember] the Sabbath day and keep it holy."[428] The first letter of the first word of this precept, zachor, is *Zayin*, indicating *Zayin*'s ability to provide the universe with a much needed taste of her cosmic energy-intelligence.

To fully understand the importance of *Zayin* in the cosmic scheme, we must examine the kabbalistic understanding of *Shabbat*. The origin of *Shabbat* is given in Genesis,[429] although the name of the day does not appear there. The Lord worked six days at creating the world, and on the seventh day He ceased His work. He blessed the seventh day and declared it holy. The special status and name of the seventh day were disclosed to Israel in the incident of the manna. The Lord provided manna for five days. The Lord did not want the Israelites to work by gathering food on the seventh day, so He instead sent a double portion on the sixth day, to last them through the seventh day. The Lord told the Israelites that the seventh day was "a Sabbath [*Shabbat*] of the Lord," which they would observe by refraining from their daily food-gathering labor.[430] The fourth utterance of the Decalogue (the Ten Utterances), generalizes the precept of the *Shabbat* and the prohibition against work on that day. The association of the *Shabbat* with Creation indicates the fundamental reasoning of *Shabbat*. The sanctity of the day is grounded in the Lord's cessation from work.

The first question that naturally arises is: How can we comprehend the concept of the "Lord's work?" Is the Lord's work in any sense similar to our own? Why, after all, should we rest on *Shabbat*? No human being has labored nearly as hard as would have been required to create the cosmos. Could the Lord's requirement of rest be based upon the fact that He toiled too long and hard at Creation?

Many religions incorporated the religious experience of the Israelites when they adopted the tradition of periodic holy days of rest. When properly understood, the observance of festivals like *Shabbat* can provide a heightened spiritual awareness to all humankind. The purpose of biblical precepts and commandments is not to tyrannically regulate behavior, but to add meaning to life by revealing and eliciting the beauty and power of Creation through the biblical code of the universe.

Religion, as it is conventionally misunderstood, seemingly does little to alleviate the problems of daily life. If anything, it appears to represent a restrictive, stifling system of behavior. Therefore, it only stands to reason that religion is widely perceived to be an impediment to the enrichment of one's life. This is why the vast majority of humankind derives scant benefit from biblical precepts. Nor is it a surprise that the *Shabbat*, as it is commonly perceived and practiced, contributes little or nothing toward the elimination of man's inhumanity to man.

The kabbalistic worldview seeks to reinterpret the significance of the festivals and make their meaning relevant and experiential. Similarly, the Passover story of the Redemption from Egypt is an occasion for meditating upon humanity's relationship with the cosmos.[431] The Passover festival directs our attention to the need to draw upon the Lightforce of the Lord's energy-intelligence.

The sages were eloquent on the value of observing *Shabbat*: "If Israel keeps two *Shabbats* as it should be kept, the Messiah will come." They say that *Shabbat* is equal to all the other precepts of the Bible.[432] The Lord said to Moses, "Moses, I have a precious gift in My treasury whose name is *Shabbat*, and I want to give it to Israel."[433]

Included in the biblical description of the six days of Creation is the passage: "And there was evening and there was morning."[434] This verse is omitted for the seventh day, the *Shabbat*, where the Bible says: "And on the Seventh Day, the Lord finished His work which He had made." Thus, the Scriptures acknowledge that the Lord did work on the seventh day, but the Bible does not say what the Lord created. What seems to emerge from these considerations is that *Shabbat* has no direct connection with the concept of rest.

Kabbalah teaches that the seven days of the week are reflections of the seven original days of biblical Creation. Each day represents one of the seven *Sefirotic* energy-intelligences that govern the cosmos. The cosmic consciousness of *Shabbat* does not partake of the Desire to Receive for Oneself Alone. Man, given the opportunity to choose freely, can merge with *Shabbat*-consciousness and disconnect from the energy-intelligence of negative desire.

This is precisely what the Lord meant when He called the *Shabbat* a "precious gift." A gift indicates something we receive without effort. *Shabbat* is a gift insofar as it represents a flow of energy without any restriction. On *Shabbat*, the universe experiences an endless flow of energy, and desires are fulfilled without the necessary restriction. By contrast, the other six days are filled with uphill battles between opposing forces, with humankind acting as the balancing factor.

Realistically speaking, the majority of people are not in accord with *Shabbat*, nor do they experience it as a precious gift. For many, observance of *Shabbat* does not provide the spiritual elevation that such a gift should provide. In Kabbalah, the observance of *Shabbat*, with the aid of methodical meditation, stimulates the harmonious movement of pure thought-intelligence, resulting in a lingering sensation akin to that of listening to the most exquisite musical harmony.

The list of work forbidden on *Shabbat* by traditional Judaism encompasses only those activities that stimulate negative energy. Thus, the labor of a waiter—whose job, by its very nature, consists of serving others—is not proscribed, but if the air conditioner were accidentally turned off on a hot, sticky *Shabbat* in August, turning it on again would violate the *Shabbat*, even though little "labor" is involved.

The *Zohar* treats the *Shabbat* at length because *Shabbat* is a time when the entire arrangement of the cosmic order is altered. The terms "evening" and "day" symbolize the basic intelligent energies of positive and negative. The six days of Creation consist of a constant struggle between Right (positive) and Left (negative) Column energy. The awesome task of creating harmony between the two is left to humankind. For most, the job is taxing. Unification between these cosmic forces results from our efforts to restrict our Desire to Receive for Oneself Alone. If we fail to harness the Left Column, we are cut off from the flow of energy. The six days of the week give us a used-up feeling. On *Shabbat*, however, the structure of the cosmos undergoes a dramatic change. The energy-intelligence of the Left Column, the Desire to Receive, is placed in a motionless state of inactivity.

During the six-day period, the internal energy-intelligence of the Left Column provides a necessary link to the circuit of energy. As with the case of the electric bulb, the negative pole is united with its positive counterpart by the filament that provides the necessary restrictive energy-intelligence to form a circuit of energy. If the filament fails in its function, the circuit is broken and energy ceases to flow. Similarly, humanity must maintain an activity of restriction, a Central Column consciousness to keep the circuit open. Otherwise, the flow of energy comes to a halt and humankind becomes drained.

On *Shabbat*, however, we are free of this responsibility. The cosmos was created with an inbuilt, ongoing restrictive process that keeps the circuit of energy open without human intervention or participation. This is indeed a gift. The doctrine of Bread of Shame is deemed nonexistent on *Shabbat*, and the Desire to Receive for Oneself Alone does not exercise a limiting effect within the cosmos. On *Shabbat*, one can receive endlessly without fear of causing a short circuit. Put another way, we may say that on *Shabbat*, the filament will never burn out.[435]

There is, however, one condition that might upset the *Shabbat* function of the cosmos. Although the Cosmic Consciousness of *Shabbat* exists outside the nature of the Desire to Receive for Oneself Alone, man—given the opportunity of free choice—can choose to absent himself from the *Shabbat* framework by acting to gratify the Desire to Receive for Oneself Alone. The electron, whose internal energy-intelligence is the Desire to Receive for Oneself Alone, is cosmically inactive on *Shabbat*. Man alone has the power to alter this state of consciousness. He can reactivate the negative energy-intelligence, thereby increasing the need for Central Column energy. Then the automatic flow of energy is no longer assured, man's needs are no longer fulfilled, and the mad race is on again. Unsatiated humans, with a drive to achieve the things they believe will bring them a life of peace and tranquility, flood the universe. Everyone seems to have what someone else needs. No one has enough. There just doesn't seem to be enough energy to go around.

The *Zohar*'s decoding of the biblical Creation narrative introduces a total reinterpretation of *Shabbat* that involves no connection between *Shabbat* and physical rest. Work, as understood by the *Zohar*,[436] revolves around the frenzied activity of negative energy-intelligence, the Desire to Receive for Oneself Alone. The true meaning of rest is the Desire to Receive being in a state of fulfillment, free from stress. The *Zohar* emphasizes the importance of avoiding any contact with the Desire to Receive for Oneself Alone on *Shabbat*. *Shabbat* allows us to deactivate the central role of the Desire to Receive, permitting us to set aside, for a time, the illusion of fragmentation that hides the true nature of existence, unity.

Stress is an essential aspect of life. The ongoing interaction between organism and environment often involves a temporary loss of flexibility. But these phases of trauma and imbalance are transitory and exist only so long as an interruption of energy has taken place. A circuitous flow of energy restores balance and can even transcend space and time. Cosmic *Zayin*'s recognition of the role of stress in the creative process led her to plead her suitability as the channel for Creation. The energy-intelligence she manifests is considered holy because it represents a constant circuitous flow of energy.[437] Under her sway, the universe would be assured of attaining *menucha* (tranquility), which would result in the negative energy-intelligence of the *klipot* (the Desire to Receive for Oneself Alone) being laid to rest, and the difficulty in maintaining restriction, the manifestation of the Central Column, being solved.

When the Desire to Receive becomes neutralized by cosmic *Zayin*, she becomes a "crown upon the head of *Zeir Anpin*," and the outer-space connection is achieved.[438] Thus, the universe once again returns to the tranquil state of Adam-consciousness that prevailed before the original sin. This is the power of the *Shabbat*. However, this altered state is only temporary because the World of Action, the terrestrial level, has yet to undergo the all-embracing elevation of consciousness that will take place at the time of the *Gemar HaTikkun* (Final Correction).

Each energy-intelligence of the Desire to Receive may achieve its correction and connection to *Zeir Anpin* only when its Central Column becomes manifest. So long as all its parts retain an element of separateness, the six-day war, begun at the time of the original Creation, will rage on. Only when each and every energy-intelligence has given up all pretense of isolation will the World of Action, with its inherent consciousness of Desire to Receive for Oneself Alone, surrender its power and allow us to freely transcend the evil *klipot*.

The sages tell us that we will know when all souls have achieved their *tikkun*, their connection to *Zeir Anpin*. It is said that this will occur "when the sun and moon shall be of equal brilliance."[439] No longer will positive and negative retain their individual energy-intelligences. Rather, they shall reunite as interrelated parts of the unified whole, and the six-day battle between day and night will end. Then, peace between Heaven and Earth will be restored at last. Until that day, there will continue to be a six-day cycle of activity followed by a *Shabbat* of rest. This condition will endure until the Final Correction of the universe, at which time the eternal tranquility of *Shabbat* will be revealed.

"Your energy-intelligence, which seeks to rid the universe of greed and corruption through complete rest, is as yet incomplete," the Lord told cosmic *Zayin*. Her weekly recycling required man to wage war with the Dark Lord for six days before achieving *Shabbat*, the energy-intelligence of rest. Cosmic *Zayin* was an integral energy-intelligence of the Lord's battle station, but the paradox remained. Not only did she incorporate the energy-intelligence of rest and tranquility; at the same time, she was the channel for the energy that had reduced the Earth to a condition of hardship and despair.

The paradoxical nature of *Zayin* is not alone in the cosmos. Scientists are confronted with paradoxes every day. The kabbalistic worldview explains the reason for—and necessity of—paradox. If humankind represents and initiates the cosmic rhythm of our universe, then paradox stems from the dual nature of man. The positive activity of humankind infuses the cosmos with rest and tranquility. The negative energy-intelligence of the Desire to Receive for Oneself Alone creates a cosmos of fragmentation and chaos. Cosmic *Zayin* is the channel for physical expression of both of these tendencies. The determinator of the expression of these tendencies is humankind.

If cosmic *Zayin* can be seen as a double-edged sword, the likeness of the double-edged sword is also an apt symbol for

Malchut.[440] During the six days of the week, *Malchut* resides with *Netzach* (Victory) of cosmic *Zeir Anpin*, the outer-space connection. This implies that during the week, *Malchut* becomes a "sharpened sword" to ward off the attacking *klipot*. *Rumheh dekroveh* (battle spear) is an attribute of *Zeir Anpin*. Resembling the letter *Vav* of the Tetragrammaton, the spear is used to pierce the *klipot* of the Dark Lord.

The Lord concluded His reply to *Zayin* by saying, "On *Shabbat*, you are a crown upon and over the head of *Zeir Anpin*, but the dominion of that state of consciousness is only temporary. Consequently, your channel can also serve as an energy-intelligence of warfare and holocaust. The universe requires a channel that can assist humankind to fare better in their effort to achieve the *Gemar HaTikkun*." The Lord, therefore, rejected the plea of cosmic *Zayin*, for she represented the doctrine of paradox. The Hebrew translation of the pronounced letter *Zayin* is "war," a far cry from the peace and tranquility she also encapsulates.

With head lowered, cosmic *Zayin* departed from the stage of the creative process.

State of Mind;
Mind and Health;
Big Bang;
Physics;
Quantum;
Cause and Effect;
Wireless Broadcasting;
Dark Lord;
Thought;
Desire to Receive;
Plus and Minus;
Power of *Alef Bet*;
Time, Space, and Motion

CHAPTER 19

THE LETTERS
VAV AND *HEI*

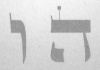

GOD USED BEAUTIFUL
MATHEMATICS IN CREATING
THE WORLD.

—PAUL DIRAC

robing the future of the high-tech frontier, the letter-energies beheld a dreadful vision of the Dark Lord. Flashing like bolts of black lightning across the endless cosmos, the Dark Lord left darkness, death, and desolation in his wake. This did not discourage the letter-energies, nor did it dampen their passion for the battle to come. Each stood ready to counter Satan with a defense that she believed would banish the Dark Lord's negativity forever. Each felt worthy of the task of defending mankind and the universe against the forces of evil. Yet no matter how high their confidence may have been or how firm their convictions, the essential question remained: Which letter of the *Alef Bet* would be chosen as the channel for the Lord's Creation, and why?

The time was near for cosmic *Vav* to present her claim. Examining the pleas of the earlier letter-energies and the reasons for their rejection, *Vav* found herself hard-pressed to articulate a compelling reason why she should be chosen as the channel for Creation. Fear did not cause her to hesitate. As one of the letter-energies of the Tetragrammaton יְ־הֹ־וֹ־הֹ, the four-letter symbol of the ineffable name of the Lord, *Vav* possessed power equal to or greater than any of the other letters of the *Alef Bet*. In a fair fight with the Dark Lord, *Vav* felt certain that she would handle herself admirably and would ultimately fare well. Yet many letter-energies of considerable merit had come before the Lord already, only to be turned away. Why, then, should she be chosen? What, if anything, had the others failed to comprehend? How should she approach the Lord? What should constitute the essence of her plea?

As if the universe did not have enough problems, it now seemed certain that violence would be established as a permanent fixture in the human psyche. With the Desire to Receive for Oneself Alone as humanity's motivating impulse, even one person's aggressive instincts could be a dangerous force on Earth. Sooner or later, humankind would have to discover a humane way of dealing with an endless storm of violence and destruction.

Vav's powers of prophecy allowed her to glimpse a future in which humanity would have access to weapons capable of global destruction. She saw a world in which animosity and hatred—the by-products of humanity's violent impulses— would pale in comparison with nuclear fallout. Why, *Vav* wondered, would the Lord give man the motivation and the wherewithal to arm himself with enough explosive power to destroy the world ten thousand times over?

The problem, it seemed to *Vav*, was to find a way of preventing violent behavior from becoming established within the cosmos. One approach would be to limit the amount of violence to which humankind would be exposed. The chosen channel for the all-embracing unified Lightforce would hopefully be in a position to prevent the destruction of the universe by serving as a strategic deterrent against Satan, while at the same time permitting man access to the infinite Lightforce of the Lord.

Despite the Lord's convincing demonstrations of the vulnerabilities of the earlier letters to an attack by the Dark Lord, cosmic *Vav* still considered herself to be a fitting Vessel to act as the channel for Creation. As for the attribute that would qualify her to serve in that exalted capacity, it would have to be her membership in the most elite and powerful force in the Lord's corps of intelligent energies: the Tetragrammaton. With this in mind, she approached the Throne of the Lord and said, "O Lord of the Universe, may it please You to accept my role as the channel in the Creation of the world, inasmuch as I am one of the four cosmic energy-intelligences that makes up the Supreme Tetragrammaton."[441]

The plea of cosmic *Vav* depended heavily upon the exact meaning of the Lord's rejection of cosmic *Yud* as the channel for the creative process, for *Yud*, too, was a member of the Tetragrammaton. If the Lord's vision of a cosmic paradise required an energy-intelligence that would permanently remove the Dark Lord from the cosmic stage of activity, then the power of *Vav* would probably be more than sufficient to achieve this objective. However, as long as the balance of power depended upon human activity, it seemed certain that the Lord would ensure the Dark Lord's survival. Concerning the future, only one thing seemed certain, and that was its uncertainty.

Even the most powerful energy-intelligences of the *Alef Bet* had to concede that no known or foreseeable force could guarantee the destruction of every negative energy-intelligence launched by Satan. Some part of the Dark Lord's attack force would penetrate any security shield, causing death and devastation beyond imagination.

Perfection, then, did not seem to be the Lord's principal criterion for choosing a channel for Creation. Nor would sheer destructive power secure that preeminent post for a letter-energy. Slowly, it began to dawn on the remaining letters that perhaps what the Lord had in mind was not a devastating offense but a strong defense. Perhaps it was time to accept the possibility that rather than obliterating the Dark Lord, they would have to be satisfied with a system that would nullify enough of the *klipot* to permit humanity the opportunity of repentance and survival. But could a deterrent system on such a limited scale still be effective?

Cosmic *Vav* harbored no doubts regarding her ability to act as an effective alternative to man's self-inflicted destruction. The acknowledged threat to world and cosmic stability made cosmic *Vav*'s internal intelligent energy appear to show the most promise of all the letter-energies who had come before the Lord, for *Vav* harbored a vision of a world unified by her energy-intelligence. She was eager to unleash the energy of the Central Column.

The Bible says: "And it came to pass, as soon as he [Moses] came close to the camp, that he saw the calf, and the dancing; and Moses' anger waxed hot, and he cast the Tablets out of his hands and broke them beneath the mount."[442] The breaking of the Tablets precipitated the destruction of both the First and the Second Temples, for the internal energy-intelligence of the First Temple was drawn from both sets of Tablets, the Left and the Right, while the Second Temple drew energy only from the Left.

"Why," asks the *Zohar*, "did the Tablets fall and break?" They broke because the internal energy of the letter *Vav* flew out of them and vanished. This power of *Vav* is hinted at in the verse: "And the Lord formed man of the dust of the ground, and breathed into his nostrils the breath of life; and man became a living soul."[443] The *Vav* of the Hebrew word *vayitzer*, which means "formed," encapsulates the Tree of Life. When Israel corrected the sin of Adam, the Israelites were deserving of the energy-intelligence of *Vav*, which was designed to create freedom from a world of instability. In this way, the Israelites earned liberation from the Dark Lord. However, the sin of the golden calf undid their correction, and the Tree of Life became concealed. In its place appeared the Tree of Knowledge of Good and Evil. Israel then received the energy-intelligence missing from *Vav*'s Tree of Life, which consisted of life stemming from its Right side and death emanating from its Left.

This is what the sages meant in reference to the new Tablets that Moses presented to the Israelites after the golden calf incident. "For those to the Right, the Bible was a potion of life; from the Left, a deadly poison."[444] Therefore, the sage Rav Akiva cautioned his students to be conscious of the separation and disorder that returned to the cosmos.[445]

What seems to emerge from the *Zohar* is the feasibility of cosmic *Vav* maintaining universal stability and an orderly cosmos. The advantage Satan appeared to maintain over corporeal man and his physical universe could shift dramatically, as the enormous spiritual vitality released by an infusion of cosmic *Vav*'s energy-intelligence moved the universe towards order and unity.

As part of the Tetragrammaton, cosmic *Vav* had power and cunning sufficient to defeat even the most deceptive measures the Dark Lord might unleash against humankind. *Vav* could provide exactly the kind of cosmic support that would give Earth's inhabitants a fighting chance. The Tetragrammaton represented the highest, most potent state of cosmic energy in the universe. This was cosmic *Vav*'s secret weapon. Hence, she asserted, "My dimension of Light and energy-intelligence is capable of providing the grand unification for which the universe will one day desperately yearn."

As we've seen, *Yud* is also a part of the Tetragrammaton, but her plea had been denied because she extended beyond the protection of cosmic *Shin*'s protective shield, making *Yud* vulnerable to attack by Satan. This vulnerability might have given the Dark Lord the opportunity of seizing rulership over all of cosmic space and preventing *Yud* from regaining her place within the mighty Tetragrammaton. If this were to happen, all would have been lost.[446] Cosmic *Yud*'s presentation and its denial took place at a cosmic conference convened by the Lord Himself, with cosmic *Vav* and all the other letters in attendance.[447] Despite the enthusiasm of cosmic *Yud*'s plea on her own behalf, her appeal was denied because, paradoxically, her position within the awesome power structure of the Tetragrammaton was simultaneously her greatest strength and her greatest weakness. All those present sensed the immensity of *Yud*'s energy-intelligence and knew that she was capable of defending the portals of universal space against any Dark empire. With the four-letter Name of the Lord as her weapon, it seemed certain that space could remain a peaceful sanctuary for all humankind. But it was not to be.

The Lord said to *Yud*, "You are engraved within Me, marked within Me. My Desire and Energy-intelligence is in you. Consequently, you are not the suitable cosmic channel for Creation."[448]

Vav was also a component of the Tetragrammaton. If *Yud*'s presence within the Tetragrammaton was the Lord's reason for rejecting her as the channel for Creation, then what made *Vav* believe a similar fate would not befall her? Cosmic *Vav* reasoned that *Yud* was rejected because *Yud* was the brain of the Tetragrammaton system, and if she became corrupted, there would be no hope for a universe free from the threat of self-destruction.[449] If the Dark Lord succeeded against *Yud*, the entire Tetragrammaton system would fail. However, if cosmic *Vav* were compromised by the Dark Lord, the Tetragrammaton's retaliatory capacity would remain intact.

Vav mistakenly believed that her energy-intelligence, combined with that of cosmic *Hei*, could combat the growing dangers facing mankind. *Vav* did not understand that she and *Hei* were as important to the maintenance of the Tetragrammaton as cosmic *Yud*; thus, their defeat would have the same apocalyptic ramifications as *Yud*'s defeat. Were *Vav* to be used in the Creation of the world, she would be equally vulnerable to corruption.

The Lord's reply to *Vav* was: "The Tetragrammaton system in its entirety requires the protective custody of the security shield of the *Shin*. You, *Vav*, must not venture beyond *Shin*'s perimeter. Whatever the outcome, the Tetragrammaton must stay clear of any contact with the *klipot* of the Dark Lord."[450]

Cosmic *Vav*'s venture into the cosmic arena would turn outer space into an energy-intelligent battlefield, adding dangerous complexities to the already precarious position of man on Earth. The overriding issue was the need to preserve the incorruptibility of the Tetragrammaton system. If the universe were indeed turned into a battlefield, the Tetragrammaton would provide the only hope for preserving the galaxies and the righteous few.

Continuity was the primary thought that hung over the universe and the human species. The Lord had stripped the issue down to a single question: Should the universe, with all its inhabitants, be permitted to come to an end without the *tikkun* process? The reply of the Lord silenced whatever other thoughts cosmic *Vav* might have had. It was certain that the Lord meant to protect and preserve the Tetragrammaton at all costs, thus ensuring that mankind might have an opportunity to complete their *tikkun* and achieve the Final Correction.

Disappointed, *Vav* stepped down from the cosmic arena of selection.

Cosmic *Hei* no longer felt the need to enter a plea to serve as the channel for the creative process. She, too, occupied an exalted position within the Tetragrammaton system. The Lord's reply to *Vav* made *Hei* realize that her energy-intelligence was also a vital link in the grand unification of the universe. The young universe could ill afford to gamble with its one certain defense system: the all-powerful Tetragrammaton. Both *Hei* and *Vav* were needed to operate this important battle station and ultimately to bring humanity to accept the Lord's vision for a stable universe. The Lord's plan required humanity to make world harmony and order a central goal. For this purpose, the Tetragrammaton provided the only hope.

Cosmic *Hei*, the Lord's obedient servant, had no choice but to also step down and permit the process of selection to continue.

Consciousness;
Antimatter;
High and Low Energy Patterns;
Why the Earth Was Created;
Genetic Code;
Human Mind;
The Rising Flame;
Consciousness;
Blue Light;
Process of Give and Take;
Assisting the Poor;
Big Crunch;
Fall of Adam;
Spacecraft;
Metaphysical Space Systems

CHAPTER 20

THE LETTERS
DALET AND *GIMEL*

NO POINT IS MORE CENTRAL
THAN THIS, THAT EMPTY
SPACE IS NOT EMPTY.
IT IS THE SEAT OF THE MOST
VIOLENT PHYSICS.

—JOHN A. WHEELER

magine the universe is a board game and our objective in the game—call it Universal Consciousness—is to secure enough antimatter to bring about the grand unification of the universe. Antimatter is valuable because it preserves universal balance and symmetry, thus serving the spiritual needs of the universal community. But antimatter has a tendency to vaporize matter; the very substance that the Technocracy is struggling desperately to accumulate. Hence, it is in the best interests of the Technocracy and the forces of Materialism to rid the universe of antimatter. The accumulation of matter fuels greed and insensitivity, which are exactly the attitudes needed to embark upon the aggressive campaign of space exploration and technological innovation that will accelerate the further accumulation of matter, thus locking the Technocracy into an ever-widening spiral. Vast regions of nether space are completely devoid of antimatter as a result of the Technocracy's efforts, and it is up to us, we who battle bravely in the name of Universal Consciousness, to locate the remaining antimatter, protect it, and disperse it to those parts of the universe where it is needed.

In recent years, there has been an explosion of interest in the scientific community concerning the elusive, ever-present substance called antimatter. Can it be that there are antimatter extraterrestrials, and even antimatter galaxies, existing in other parts of the universe? Physicists have suggested the possibility that invisible antimatter could account for as much as ninety percent of the matter in the universe. A better understanding of antimatter, scientists claim, might provide insights into the cosmic glue known as the "strong force," which binds together the nucleus of the atom.

It is jarring to consider that the vast preponderance of matter in the universe might be antimatter. If this is true, how is it that we find so little evidence of antimatter in our neck of the universe? Logically, if ninety percent of all matter is antimatter, it stands to reason that evidence of antimatter would not be so hard to find. Perhaps, in trying to solve the mystery of the missing ghost matter, we might scrutinize the possible reasons for the existence of antimatter. The startling possibility of a fundamental cosmic connection between man and the universe is one of many astonishing concepts that the *Zohar* boldly advances:

> *One night when Rav Isaac and Rav Yehuda were sitting up studying the Bible, the former said, "The Kabbalah teaches us that when the Lord created the world, He created the Lower World after the pattern of the Upper World and made the two the counterparts of each other. His energy-intelligence, therefore, must be both Above and Below."*

> *Rav Yehuda replied, "Assuredly this is so, and that He created man to be superior to all. This is indicated by the verse: 'I have made the Earth and created man upon it.'[451] The sages understood the true meaning of this verse. In their wisdom, they translated it as: 'I have made the Earth for the purpose of creating man upon it.' I say that cosmic unity depends upon man to complete the organic whole."[452]*

For all but a few scientists, the ideas presented by the *Zohar* represent scientific heresy. Nevertheless, science provides no definitive explanation for nature's infliction of havoc and destruction upon the galaxies. Despite science's carefully maintained image as a source of immutable truth, most scientific theories are based on fallible interpretations of data. For all their pretension to unassailable truth, few scientific conclusions are immune to uncertainty. The question remains: Who or what lies behind all negative cosmic activity? The *Zohar* declares that the culprit is humankind.

Kabbalists restate and defend the ancient belief that human beings consist of a mysterious combination of physical matter and intangible spiritual substance. The physical uniqueness of the human form is derived from a genetic code, but the individual's inner nature is a result of Divine creation. The quality that sets each person apart from all others is the nonmaterial self that enters the body during its embryological development or at the time of birth, stays with the body all of its life, and survives after the body's physical death. This "ghost in the body" is responsible for everything that makes us distinctively human.

The human mind has the ability to grasp realities higher than those conceived by conventional science. In its restless need to express itself, the mind only appears to arbitrarily cut the seamless fabric of reality into myriad patchwork segments. In an altered state of consciousness, the mind is capable of restoring the cloth of Creation to its original pristine condition.

Cosmic *Dalet* and cosmic *Gimel*, having previewed the drama of Creation, made their entrance together onto the cosmic stage to state their case before the Lord as to why they should be the channels for Creation. They knew that individually their intelligent energy-forces were insufficient to enforce the grand unification of the galaxies, but together they represented an energy-intelligence of substantial power. One glance at *Dalet* provides immediate evidence for why the Lord might deny her if she made her plea alone. The Hebrew word *dal* דַּל meaning "poor," spells it out at once: Cosmic *Dalet*'s internal energy epitomizes impoverishment.

> *The Zohar says: They now rose to depart, but Rav Shimon said, 'I have still one thing more to tell you. It says, "For the Lord is a consuming fire,"*[453] *and in another place, "You that cleave to the Lord are all of you alive this day."'"*[454] *The apparent contradiction between these two lines has been discussed at length among scholars. These discussions have established that there is a fire that consumes fire and destroys it, one manifestation of fire being stronger than the other. Pursuing this idea, we may say that he who desires to penetrate the mystery of the all-embracing unity should examine the flame that rises from a burning candle. A flame can only rise from something physical. In the flame itself, there are two lights: the upper one is white, the lower blue. The white light, the higher of the two, reaches upward while seeming to rest on the blue as if upon a throne or a pedestal. The two are inseparably connected, the white resting and being enthroned upon the blue. The blue base, in turn, is attached to the wick, which is attached to the wax beneath it, which feeds the flame and impels it to hold up and cling to the white light*

> *above. The blue light sometimes turns red, but*
> *the white light above never changes color.*

Kabbalists interpret this to mean that the lower energy-intelligence consumes anything that is below it, but the higher intelligence does not consume that which is beneath it. Thus, the energy-intelligence of blue or black (the effect of combustion) is associated with destruction and death. This is why Moses said, "For the Lord your God is a consuming fire," meaning literally, He is capable of consuming all that is beneath Him. This is why Moses also said, "Your Lord and not our Lord." Moses was in the "white" region of consciousness, which does not consume or destroy. The white light's energy-intelligence is symbolic of the Lightforce of the Lord. Thus, the ascending flame symbolizes the highest mysteries of wisdom.

> *Rav Shimon continued: "The second Hei of the*
> *Tetragrammaton is the blue or black light, which*
> *is attached to the Yud, Hei, Vav, which are the*
> *white lights of energy-intelligence. Sometimes,*
> *this blue light is not the letter Hei but contains*
> *the energy-intelligence of the letter Dalet. That is*
> *to say, when Israel does not cleave to or connect*
> *to the internal energy-intelligence of the white*
> *light from Below so as to make the blue light*
> *burn and cling to the white light, then the*
> *energy-intelligence of the blue light is a source of*
> *destruction. It is Dalet. However, when Israel is*
> *connected to the internal energy-intelligence of*
> *the white light [the sharing consciousness], the*
> *blue light is then considered. Where male and*
> *female [the negative energy-intelligence of the*

> Desire to Receive] are not united and unified as
> a one basic whole, the letter Hei is eliminated
> and only the energy-intelligence of the letter
> Dalet remains."[455]

Dalet's potential for aiding in the establishment of the grand unification was obvious. In essence, without her internal negative energy-intelligence, the Lightforce could not manifest in the cosmos, a situation comparable to a flame being unable to manifest without a candle. Nevertheless, cosmic Dalet was fully aware that her own particular energy-intelligence was "impoverished." She did not act as an energy-intelligence channel for the Lightforce of the Lord. Nor did cosmic Gimel possess the power to act alone as a channel for Creation. Consequently, they decided to approach the concept of Creation from a fresh perspective. No letter-energies had yet contemplated the idea of offering a dual channel for the creative process.

Cosmic Dalet and cosmic Gimel began with the accurate assumption that symmetry would be an essential property of the World to Come because any unified cosmology would exhibit a dual nature of opposing forces. As manifestations of the two fundamental forces in the creative mechanism, they considered themselves capable of bringing an end to fragmentation and disunity. United, they could assist humanity to reach the ultimate goal of completing the tikkun process, which they believed would automatically nullify the power of the Dark Lord.

The universe is actually much simpler than outward appearances might lead us to believe. In fact, everything in existence can be reduced to an absurdly simple formula: give and take. Oriental philosophy describes this phenomenon in terms of the male principle, Yang, and the female principle, Yin. Science calls it positive and negative polarities. Kabbalah calls it Desire to Receive for the Sake of Imparting and Desire to Receive for Oneself Alone. Cosmic *Gimel* and cosmic *Dalet* symbolize this unity of opposites: *Gimel* representing the positive, giving aspect, and *Dalet* symbolizing the negative dimension of receiving.

Actually, in the grand scheme of things, *Dalet* and *Gimel* were not distinct entities at all, but merely separate manifestations of the same underlying interaction—the all-embracing unified whole—which is precisely why *Dalet* and *Gimel* perceived themselves as a perfect duo for the creative process, and why they felt confident that together their unified energy-intelligence was capable of initiating the great unification.

While cosmic *Dalet* ד —composed of the three Hebrew letters *Dalet*, *Lamed*, and *Tav* דלת —indicates impoverishment on her own, when she is united with cosmic *Gimel*, the vowel "a" of *Dalet* becomes transformed to the vowel "e," and she is now considered a *delet* (a door, a gateway) to the two great Lights or energy-intelligences that make up the grand unification: the Lights of Mercy and Wisdom.

"Cosmic *Dalet* is considered to be the *Delet* of the Holy City of Jerusalem. When the energy-intelligence of Mercy, which is the *Gimel*, the doer of good deeds, becomes united with *Dalet*, symbolic of impoverishment, *Dalet* then undergoes a transformation and becomes the gateway and door to both vital energies of the unification process: the Light of Mercy[456] and the Light of Wisdom."[457] This concept of the *Dalet* acting as a gateway is stated in the verse:[458] "Open to me the gates of righteousness."[459] The *Talmud* emphasizes this idea when it says: "The shape of the letter *Gimel* ﬡ in biblical script has the foot pointing forwards."[460] This implies that everyone should step forward and assist the poor. The *Dalet* ﬢ, on the other hand, has its upper section extended outward, indicating the necessity for the poor to permit themselves to be available for the charity dispensed by the well-to-do. Each complements the other, with no distinction between the two. There can be no sharing without an energy-intelligence of receiving.[461] If the energy-intelligences of the individual cosmic letter-energies failed to meet the Lord's requirements, *Dalet* and *Gimel* reasoned that perhaps the galaxies might fare better if two of the letters joined forces in a dual effort to defeat the Dark Lord.

Some physicists make the bleak prediction that the world will end in an implosion, a Big Crunch, the antithesis of the Big Bang. However, this prediction assumes that there is nothing special about man's place in the cosmos. The *Zohar* states otherwise,[462] telling us that invisible matter, which consists of pure energy-intelligence, will warn us if the universe is on the brink of collapse. The thought-activity of humankind, in the final analysis, will determine if we survive or not. Depending on the level of consciousness to which we have ascended (or descended), civilization may either be destroyed in a fiery apocalypse that will dwarf anything ever imagined in science fiction or our galaxy may be spared and never have to face up to the prophecy of cataclysm.

Cosmic *Dalet* and cosmic *Gimel* were certain that together they could provide sufficient energy-intelligence to prevent the overthrow of the universe through a Big Crunch or any other catastrophic assault by the Dark Lord. So they approached the Lord and said, "Hear our plea, Lord, for we are most desirous of being chosen as the dual channel for Creation. Together we represent the power of balance and symmetry. Thus, we could generate enough Central Column energy to thwart any massive attack on the cosmos by Satan."

Although cosmic *Dalet* was located well within the Lord's security shield, she still possessed a vulnerability that Satan could exploit to gain access to her internal force. *Dalet*'s problem was similar to that of cosmic *Shin*. The corner of the roof of cosmic *Dalet*, protruding with the Light of *Chesed* (Mercy),[463] gave Satan an opportunity to effect a breach in *Dalet*'s connection with cosmic *Gimel*. If this were to happen, *Dalet*'s link with the Lightforce (Light of Wisdom) could become severed, and Satan could neutralize her internal energy-intelligence. Then the Dark Lord would be able to transform the truncated *Dalet* ד into a likeness of *Resh* ר (poverty). *Dalet*'s capture by Satan would thus give the Dark Lord a powerful new spacecraft to add to his fleet.

Before the fall of Adam and the subsequent shattering of the Vessels, *Kuf* and *Resh* maintained a perfect unification and symmetry of interactions. Only after the first disruption within our universe did the Dark Lord find his opportunity to manifest severely negative intelligence. The balance of power rested on the side of the Dark Lord, however, were the universe to be created through *Dalet*, there would always be a danger of Satan seizing ultimate control. Empty space could be reduced to a vast wasteland of static negative intelligence, triggering calamities beyond measure throughout our world.

Although geologists comprehend plate tectonics better than any other aspect of the phenomenon of earthquakes, the greatest mystery is why earthquakes happen at all. The key to solving this mystery lies in isolating the cause of the geologic imbalance that makes the Earth's crust change position. Seismologists, however, have no way of knowing precisely when one geologic plate will break free of another and set the Earth's crust trembling. According to the *Zohar*,[464] the answer is to be found in the hot, turbulent atmosphere of *Malchut*, the illusory space where the Lightforce of the Lord, supported by positive human activity, wages battle with the Dark Lord, assisted by negative human activity.

Cosmic *Dalet* makes herself available to persons of means as a conduit for the transfer of energy from the rich to the poor, thus making possible the establishment of a complete circuit of energy. Without her cosmic function, humanity could not make manifest the internal energy-intelligence of benevolence, for without the impoverished, the rich could not genuinely share. If *Dalet* became poorer (*Resh*) by being deprived of the Light of *Chesed* (Mercy), the opportunities for chaos and disorder would be vastly increased, posing a threat to the very balance of the universe.

"My cosmic-letter armada could ill afford the loss of any lettered spacecraft, especially one so valuable an ally as you, *Dalet*. You and *Gimel* must remain side by side," said the Lord, "since it is written: 'The poor shall not cease from the land.'[465] Stay side by side together, as you both need to take extraordinary precautions to avoid seizure by the Dark Lord."[466]

By virtue of their internal energy-intelligences, cosmic *Dalet* and cosmic *Gimel* provide the metaphysical systems permitting earthbound humanity, through the complementary activities of receiving and sharing, to maintain balance and harmony in the cosmos. "As Above, so Below."[467]

Together, cosmic *Dalet* and cosmic *Gimel* departed from the presence of the Lord.

CHAPTER 21

THE LETTER
BET

STILL THERE ARE MOMENTS
WHEN ONE FEELS FREE FROM
ONE'S OWN IDENTIFICATION
WITH HUMAN LIMITATIONS
AND INADEQUACIES. AT SUCH
MOMENTS, ONE IMAGINES
THAT ONE STANDS ON SOME
SPOT OF A SMALL PLANET,
GAZING IN AMAZEMENT AT
THE COLD YET PROFOUNDLY
MOVING BEAUTY OF THE
ETERNAL, THE UNFATHOMABLE.

—ALBERT EINSTEIN

"Blessing" is an elusive concept. If you have ever attempted to describe it, you have probably fumbled with words such as luck, Godsend, and happiness, and you might just as easily have been dissatisfied because blessings are all of these and none of them. Another word that might be used to describe the idea of blessing is unity. We cannot hear, taste, touch, or smell unity, and yet we can certainly feel it when it is upon us—and when it disappears. Marriage supposedly describes the ultimate unification of two people, and yet half of all marriages end in divorce. Indeed, it is often said that the surest way to end a beautiful romance is to get married. In such instances, how quickly does a seeming blessing turn into a curse?

Most would agree that blessings have a funny way of ending up on other people's doorsteps and not theirs. According to conventional wisdom, blessings seem to rain down on others, but though we thirst for their life-giving elixir, seldom does much more than a drop seem to land on us. Blessings depend upon the perspective from which they are viewed, so one person's blessing may well be another's curse. A bowl of brown rice is equivalent to a king's ransom in the eyes of a starving man, but it might be an insult to one whose palate is accustomed to gourmet food. Yet if you served a steady diet of brown rice to the man who was starving and deprived the rich man of food entirely, it would not be long before the tables turned and brown rice would fall dramatically in the estimation of the former, while rising significantly in the eyes of the latter. In the space of a few days, the blessing–curse dichotomy would completely reverse itself.

Attitude and circumstance play a part in the way we perceive blessings. One of the fundamental traits of human consciousness is our ability to perceive the ongoing presence of time. When we see two snapshots of someone—one as a baby and another as an adult—we have no trouble determining which was taken first. Time moves inexorably forward. All this seems quite obvious and even trivial until we ask: *But does it really?* Why should time not march backward, or even sideways? If kabbalists (and also a few brave theoretical physicists) are correct and space-time is part of an infinitely dimensional continuum, why can't we remember tomorrow as well as we can remember yesterday?

We don't normally remember the future, it is true, but that doesn't mean remembering the future is impossible. Many people have experienced prescience, precognition, and déjà vu. Scientifically speaking, seeing the future is of course impossible, so scientists have advanced theories that attempt to explain precognitive phenomena as being nothing more than short circuits in the brain. Why, then, should we believe the kabbalist who tells us that human consciousness has access to the past and the future?

Scientists are the first to admit they know next to nothing about the human brain. Nor can they agree on a theory that explains the forward march of time. According to some theories, time could hypothetically run in reverse if it were possible to travel beyond the speed of light, and this might well mean that we could remember next year as clearly as last. Einstein understood the true nature of time more clearly than most. In a famous letter he wrote after the death of his friend, Michele Besso, Einstein wrote to Besso's sister and said: "Michele has left this strange world before me. This is of no importance. The distinction between past, present, and future is an illusion, although a persistent one."

One of the world's leading cosmologists, Roger Penrose of Oxford University, proposes that time is purely a psychological event. If consciousness makes time move forward, consciousness can presumably also make time run in reverse. In a way, a flipped coin, for example, exists as both heads and tails. Uncertainty exists even after the coin has landed and been covered with the hand. Psychologically speaking, time expands to encompass both possibilities until the moment the coin is uncovered.

If time runs forward in an expanding situation, such as the period between the flipping of the coin and its revelation, does it then run backward at the moment of revelation, when time contracts to encompass either heads or tails (but not both)? Is time reversed at the moment of the uncovering, and does potential then become reality? At death, do we become retro-people?

Of course, all of these speculations are hopelessly confusing. And they will remain so as long as we remain ignorant of the illusory nature of physical life and the extreme limitations of so-called "rational" thinking. Only by transcending rational consciousness can we hope to unravel the mysteries of our universe and alleviate the confusion that assails us in our everyday lives.

Because of the limitations of rational thinking, we have to accept the fact that, as Werner Heisenberg phrased it, "Every word or concept—clear as it may seem to be—has only a limited range of applicability."[468] The physical universe is replete with paradox. This fact goes a long way toward providing us with an explanation of why so much confusion exists in the universe—and ultimately in our own personal lives.

Steven Weinberg, one of the world's leading theoretical physicists and co-architect of the theory of the unification of weak and electromagnetic interactions, wrote that the more comprehensible the universe seems, the more pointless it also seems. His remark is typical of many made by scientists who conclude from their extensive research that the universe seems to have no discernible purpose, and therefore must have emerged as a result of a vast, meaningless accident.

How can anyone conclude that we live in a cosmos where chaos reigns supreme? Simple observation should tell us the universe is anything but random. Just look around. Has any phenomenon yet emerged of which we may state unequivocally that it arose without meaning or purpose, or by virtue of an accident? Every advance in fundamental physics uncovers yet another facet of the universal order. Yet no matter how deeply the scientist probes the depths of space and the subatomic realm, he or she will always encounter deeper, more unfathomable mysteries. Nature is too subtle, and far too profound, to be laid bare by the scientific method.

A few physicists, like Erwin Schrodinger, admit to the confusion to which their investigations have led them. "I know not from whence I come, not whither I go, nor who I am," he wrote. Curiously, one of the sages in the *Ethics of the Fathers* (*Pirkei Avot*, a collection of ethical teachings of the rabbis of the

Mishnaic period), having surveyed this same conceptual construct, arrived at quite a different conclusion from Erwin Schrodinger. Akabya ben Mahalalel said, "Consider three things and you will not fall into the power of sin and illusion. Know from whence you came, whither you are going, and before whom you are about to give an account and reckoning."[469]

To the student of Kabbalah, Akabya's three principles represent the three intrinsic universal forces of Right, Left, and Central Column energy. These three energy-intelligences, along with the fourth force—humanity, whose innate intelligence is one of lack or need (Desire to Receive for Oneself Alone)—combine to control the universe. These four forces are manifestations of a single underlying super-Lightforce-energy that ultimately not only accounts for all activity in the universe, but also has the capacity to maintain stability throughout the cosmos, preventing chaos and illusion from becoming the universe's dominant influences. This super-Lightforce-energy, which is the Infinite Reality, lies beyond conventional physical experimentation and finite understanding. Only within a metaphysical framework can humankind be freed from the world of illusion and gain access to this infinite reality, where clarity rules supreme.

To procure this blessing, cosmic *Bet* entered the arena where one letter-energy would be chosen as the channel for Creation. Cosmic *Bet* was well aware of the flaws in the reasoning of each letter that had come before her. For every positive characteristic that each one possessed, there was an equal and opposite force within its makeup that would have created the conditions for a universe of chaos and disorder. Had any of the previous letters been chosen to act as the channel for Creation, the cosmos would have been ruled by blind chance and not by the absolute clarity that governs the real world of the infinite.

The Dark Lord would have been thrilled if the Lord had chosen to create a universe where chaos reigns supreme. But, thankfully, this was not what the Lord had in mind. His blueprint called for a design whereby humankind would be able to discern, if it so chose, between the reality of the infinite and the illusion that would be presented by finite existence. It was essential to the creative process that humankind be given a system it could use to avoid the uncertainty of a random universe. Such was the nature of cosmic *Bet* whose energy of blessing would always be available to assist the spiritually inclined in their quest for infinite consciousness. Free will and determinism would of necessity exist as distinct, though interrelated, parts of the cosmic landscape. Thus would the spectacular pattern of the impeccably ordered universal design be revealed, but only for those who understood and exercised the principle of resistance.

In this way, the stage was set for cosmic *Bet* to make her plea: "Lord of the Universe, may it please You to establish the creative process through me, for I represent the energy-intelligent force of *berachot* [blessings] because I am the first letter in the word for that coded force."[470]

Cosmic *Bet* expressed in a word her unique energy-intelligence. Her *berachot* could bathe the entire universe in the only energy capable of washing away illusion, thereby suffusing the lower terrestrial realm with the intrinsic reality of the Upper Celestial Realm. Only the energy-intelligence of *berachot* was capable of removing the illusions of corporeal existence and revealing a cosmic reality that was, is, and always will be static, timeless, and perfectly still. Cosmic *Bet*, as messenger of this intelligence, was the only letter-energy capable of expressing the unity of the all-embracing Lightforce of the Lord.

The Lord's plan for the universe required the minds of humanity to be cleansed of their intrinsic state of deception through man's voluntary resistance to this state. Cosmic *Bet* assured the Lord that her energy could pave the way for a new reality in which the old, illusory mindset would give way to the all-embracing true reality of the cosmos.

Cosmic *Bet* was the first letter to represent a serious threat to the Dark Lord. Hers was the first energy with the capability of filling the false vacuum, thus preventing the inhabitants of the physical universe from mistakenly perceiving their corporeal surroundings to be an event without a cause. Were it not for cosmic *Bet*, the idea of a random, chaotic universe might have been taken seriously. Of all the letters who had pleaded their merits before the Lord, only cosmic *Bet*'s blessing-energy could allow man to have an internal connection with the absolute certainty of the Lightforce.

Consequently, cosmic *Bet* pleaded that her energy was necessary to dispel the illusory pattern that the Dark Lord would weave into the fabric of the universe. Otherwise, were the Dark Lord to come to power, man might abdicate his duties of determination of cosmic activity and opt for the illusion of the free and easy way out, not realizing that in the long run this would be more painful.

And what was the Lord's response to cosmic *Bet*'s plea? The Lord said, "Assuredly, through your channel I will create the world. Yes, your energy-intelligence shall be at the starting point of the creative process."[471]

What is this super-force blessing-energy of cosmic *Bet* that provides clear, definable laws and principles instead of illusory uncertainty? To more fully appreciate *Bet*'s energy-intelligence, let us turn to the *Zohar*, which states that only through observation can one arrive at truthful conclusions. To many, this outlook may seem lacking in faith, but the *Zohar* and the kabbalist simply do not accept the unique energy-intelligence of cosmic *Bet* or anything else without personal verification.

The author of the *Zohar*, Rav Shimon bar Yochai, carefully considers the question of how we can be certain of our interpretation of the metaphysical attributes of cosmic *Bet*: "And for those persons who do not know, yet have a desire to understand, reflect upon that which is revealed and made manifest in the terrestrial level, and you shall know that which is concealed, inasmuch as everything (both Above and Below) is the same. For all that the Lord has created in a corporeal way has been patterned after that which is Above."[472]

Armed with this introduction, we can now turn to what is probably the most mystifying and profound development in our understanding of the origin of the universe: the emergence of the internal blessing-energy of cosmic *Bet* that gives spiritual people access to the infinite reality, while the non-spiritual sector of humankind, those who are governed by their Desire to Receive for Oneself Alone, see only fragmentation.

Rav Shimon continues:

> *Everything Below corresponds entirely to that which is Above. This is the significance of the words: "And the Lord created man in His own image; in the image of the Lord created He him."*[473] *Just as in the firmament which covers the whole universe, we behold different shapes formed by the conjunction of stars and planets to make us aware of hidden things and deep mysteries. So, too, upon the skin, which covers our body and which is, as it were, the body's firmament covering all, there are shapes and designs (stars and planets), in which the wise of heart may behold the hidden things and the deep mysteries indicated by these shapes and expressed in the human form. Concerning this it is stated*[474] *"The viewers of the heavens, the stargazers...."*[475]

The scientist may shake his head in despair at such a statement, but the *Zohar* makes it clear that "the body of man is related to our entire galaxy and universe." An in-depth analysis of the body, combined with an understanding of which planet or part of our galaxy relates to each component of man, will open new vistas in the heavens. It therefore comes as no surprise that the famed Italian Kabbalist, Shabbatai Donolo, was also a physician. His famous work *Book of Wisdom*, a commentary on the *Sefer Yetzirah* (*Book of Formation*), explains the relationships among the planets in precise detail. (Shabbatai Donolo's *Book of Remedies* also contains a great deal of material drawn from his comprehension of the Kabbalah.)

What distinguishes cosmic *Bet* from all other cosmic letter-intelligences? A good starting point for our investigation is to understand where the coded information sequences might come from and to look at the unique and peculiar design of the Hebrew letters and the way they have become manifest.

The natural impulse in thinking about intelligence in the cosmos is to start with ourselves and then attempt to work upward toward the unknown. The path of the kabbalist is to jump straight toward the concept of an all-embracing intelligence and work downward from there. In so doing, we leave aside intermediate intelligences that may be masked by the trappings of our earthly existence, a concept broached in the *Zohar*, to which we will now turn.

> *Rav Chiya said that the Oral and the Written Law together preserve mankind, as it is written: "Let us make man in our image, according to our likeness,"*[476] *where "image" refers to the masculine (code name for the metaphysical realm) and "likeness" to the feminine (code name for the physical realm), and for this reason the Bible commences with the letter Bet. Rav Isaac said: "The Bet is open on one side and closed on the other to indicate that when man desires connection with the internal energy-intelligence of the Bible, it [Bet] is open to receive him [man] and connect with him. And when a man closes his eyes to it [Bet] and walks in the other direction [non-spiritual], then it turns its closed side to him, according to the saying: 'If you leave me one day, I will leave you*

two days,'[477] *until he returns to attach himself to it, never again to abandon it."*[478]

The *Zohar* establishes cosmic *Bet* as the celestial metaphysical gateway and connection to the all-important compendium of energy-intelligence: the cosmic code, also known as the Bible. However, merely establishing cosmic *Bet* as the letter of Creation only raises more questions in the *Zohar's* attempt to provide an understanding of our universe and our participation in it. Why was cosmic *Bet* established with her particular design in the first place? And because the kabbalist always asks why, no stone shall be left unturned until the final "why" is laid to rest.

> *Rav Yehuda said, 'The Bet has two parallel lines and a third joining them. What do these signify? One, for Heaven, portrays Zeir Anpin [the outer-space connection], and one, towards Earth, portrays Malchut [the terrestrial realm]. The line joining both parallel lines is the Lord, the code name for the Sefira of Yesod, which unites and receives them." Rav Elazar said, "These three energy-intelligence forces symbolize the Three-Column energy system*[479] *by which our universe has become associated, of which the whole cosmic code, the Bible, is comprised. Cosmic Bet is the gateway and opening of the all-embracing Amen,*[480] *which leads to the inner sanctum of the Bible. Furthermore, the letter Bet is the first letter of the word bayit (house), housing the forces of the three energy-intelligences portrayed by the three columns of*

> letter Bet ⊐ . *Cosmic Bet is therefore all-*
> *inclusive of the Bible, since the Bible begins with*
> *Bet. She is therefore the healer and stabilizer of*
> *the universe.*"[481]

We must also deal with the notion of columns and inquire about their importance to the cosmic scheme. Why is cosmic *Bet's* structure so crucial to the awesome power of the Lightforce mentioned in the *Zohar*? Where did the awesome power of the Three-Column system originate? Prior to the miraculous splitting of the Red Sea, one of the most impressive cosmic upheavals in the long history of humankind, Moses drew upon the all-inclusive positive cosmic energy by tapping into the source of this energy from the Tree of Life. How did Moses connect to this awesome power without burning out? When and where in the Bible did this secret become revealed?

The secret lies in the inherent power of the seventy-two letters. Rav Shimon tells us that the *Shechinah* was in her fullness and perfection at the splitting of the Red Sea, manifesting in herself the 72 Holy Names of God according to the threefold order, namely the Three Columns or the secret of the three verses,[482] each of which contains precisely seventy-two letters. All three verses begin with the letter *Vav*,[483] and the letter *Vav* portrays the dimension of a column. The number three should immediately draw the reader's attention to the significance of the all-embracing unified power of *Chesed*, *Gevurah*, and *Tiferet*, which are the Right, Left, and Central Columns, as well as to the three components of the atom: proton, electron, and neutron.

First Verse Right Column

וַיִּסַּע מַלְאַךְ הָאֱלֹהִים הַהֹלֵךְ לִפְנֵי מַחֲנֵה יִשְׂרָאֵל
וַיֵּלֶךְ מֵאַחֲרֵיהֶם וַיִּסַּע עַמּוּד הֶעָנָן מִפְּנֵיהֶם וַיַּעֲמֹד
מֵאַחֲרֵיהֶם:

Second Verse Left Column

וַיָּבֹא בֵּין מַחֲנֵה מִצְרַיִם וּבֵין מַחֲנֵה יִשְׂרָאֵל וַיְהִי
הֶעָנָן וְהַחֹשֶׁךְ וַיָּאֶר אֶת הַלָּיְלָה וְלֹא קָרַב זֶה אֶל זֶה
כָּל הַלָּיְלָה:

Third Verse Central Column

וַיֵּט מֹשֶׁה אֶת יָדוֹ עַל הַיָּם וַיּוֹלֶךְ יהוה אֶת הַיָּם בְּרוּחַ
קָדִים עַזָּה כָּל הַלַּיְלָה וַיָּשֶׂם אֶת הַיָּם לֶחָרָבָה
וַיִּבָּקְעוּ הַמָּיִם:

Thus, cosmic *Bet* was an all-powerful energy-intelligence simply because her structure embodied the all-embracing unified whole. Her Three-Column structure enabled cosmic *Bet* to establish blessings in the universe and to bring about the unification of all the Upper and Lower Worlds.

I would like to suggest to the reader of this book that you at least "taste" this awesome power of the universe. Scan the table of the 72 Names of God below and you will discover the

Scanning directions

כהת	אכא	ללה	מהש	עלם	סיט	ילי	והו
הקם	הרי	מבה	יזל	ההע	לאו	אלד	הזי
וזו	מלה	ייי	גלך	פהל	לוו	כלי	לאו
ושר	לכב	אום	ריי	שאה	ירת	האא	נתה
ייז	רהע	ועם	אני	מנד	כוק	להו	יוו
מיה	עשל	ערי	סאל	ילה	וול	מיכ	ההה
פוי	מבה	נית	נוא	עמם	הוש	דני	והו
מוזי	ענו	יהה	ומב	מצר	הרו	ייל	נמם
מוב	היי	יבמ	ראה	וזבו	איע	מנק	דמב

Scanning directions

Lightforce inherent in the Hebrew letters of the *Alef Bet*. If you cannot read Hebrew, may I suggest that you read the code of letters anyway? I assure you that your subconscious—the real you, the 99 percent of you—will capture the essence of the cosmic code of the Lightforce. You will retain at least some portion of the internal Lightforce of this energy-intelligence—or possibly all of it, depending on your own level of spirituality and not on the degree of your religiosity. The greater the degree of your love for your neighbor, the greater shall be your connection with the Lightforce of the Creator.

At this point, you may be asking yourself: How can a simple examination of letters that I can't even read open a cosmic connection within my inner self?

Observe the checkout counter the next time you go to the supermarket. You will notice the clerk passing your packaged food items over a scanner so that a funny-looking configuration of lines faces the scanner. The scanner relays this information to a computer, which in turn instantaneously transmits the purchase price back to the cash register.

This is precisely the interrelationship that exists between our eyes and our brain. When we scan the table of the 72 Names of God, we are entering information into our mental operating system. So if you're not familiar with Hebrew letters or words, at least you have the opportunity of copying the seventy-two letter data into the personalized computer of your mind.

We have now connected with the software program of cosmic *Bet* by understanding her Three-Column structure. We have gained knowledge, and knowledge is the connection to the true reality. Information exists whether we know it or not, but knowledge is the connection between ourselves and the information. Knowing is a form of connection. Obviously, there was an act of physical intercourse between Adam and Eve, but that is not the point emphasized by the *Zohar*. The only way information can become connected with us is when we know it, thus turning mere information into knowledge.

To know cosmic *Bet*, we had to first understand why she is structured the way she is. With that knowledge, we can proceed to make contact with the internal energy-thought-intelligence of the Lightforce that enables *Bet*'s unique blessing-energy to wield so much power. The Lightforce adjusts itself to the capability and capacity of the Vessel manifesting the Lightforce. In *Bet*'s case, the Lightforce created the ultimate engine of the universe—the unified model that would maintain balance in the cosmos. Let us therefore examine what the *Zohar* tells us about the Lightforce as it became manifest within cosmic *Bet*.

"Cosmic *Bet* is the secret of *Chochmah*. She contains the mystery of *Nekudah BeHaihkalah* (Point of the Hall) for she represents *Chesed* (Mercy) of *Chochmah*. The Light of Mercy is the 'hall' or 'house' for the Light of Wisdom (the Lightforce). This is the ultimate in 'blessing' (balance), a point illustrated by the prophet Malachi in the verse: 'I shall pour unto you a blessing, that there shall not be room enough to receive it.'"[484]

The Light that "descends" from the Endless is not diffused as it slices through the cosmos. From the beginning of its journey to the end, it is not deflected from its straight course by any metaphysical fields of energy or by any of the curtains it passes through. Therefore, cosmic *Bet* stated that she was suitable to be the channel of Creation because she alone was not affected by negative forces and influences. The Dark Lord, seeking to increase his dominion over the cosmos, could not make any connection with cosmic *Bet*, and thus could not use cosmic *Bet* as a source for additional energy. Only through a lack or deficiency can Satan find an opening by which to attack its victim.[485] Cosmic *Bet* had no such lack; therefore, she was not vulnerable to the Dark Lord's nefarious stratagems.

Here for the first time, the *Zohar* provides us with a glimpse into what constitutes "blessing": Any energy force, any entity, or any man who is completely filled with energy and who displays no lack is considered blessed.

If we examine a medical X-ray, we notice that the location of the medical problem is designated by a black spot. Why a black spot? To indicate that the flow of energy is faltering and Satan has made an inroad. The lack of energy creates vulnerability and thus an opportunity for the Dark Lord to make a surprise attack. The removal of the black spot, however, does not necessarily restore the flow of energy. A state of blessing or completeness is required, not just the mere elimination of the Dark Lord.

Various biblical narratives attest to the value and efficacy of blessing, such as Noah's blessing of Shem and Japheth;[486] Isaac's blessing of Jacob and Esau;[487] and Jacob's blessing of his sons[488] and of his grandsons, Ephraim and Manasseh.[489] A father usually blesses children by laying his hands upon the head of the child and pronouncing the verse (for a boy): "May the Lord make you like Ephraim and Manasseh,"[490] followed by a priestly benediction[491] that contains a threefold arrangement, making explicit the intent of the ordained formula. This laying-on of hands has long been known as a channel for the transference of energy.

Declares the *Zohar*: "We have been taught: Whoever has attained the degree of *Chesed* (Grace) is designated 'angel of the Lord of hosts,'[492] as it is stated in the verse: 'For the priest's lips should keep knowledge, and they should seek the law at his mouth; for he is the angel of the Lord of hosts.' Wherewith did the priest merit to be called 'angel of the Lord of hosts?' Said Rav Yehuda, 'As the angel of the Lord of hosts is a priest Above, so is the priest Below an angel of the Lord of hosts.' Who is the celestial High Priest? The angel Michael, who issues from the celestial, cosmic force of *Chesed*."[493]

The priest was established as the chariot, or link, between the *Sefira* of *Chesed* and the terrestrial realm. By virtue of the priest's embodiment of the Lightforce of *Chesed*, he was considered the channel for blessing. The priest was the connection to the Lightforce, the all-embracing unified whole, and therefore was endowed with the awesome power of the Lightforce to bless mankind. The secret was the Light of Mercy, and this was the combination that cosmic *Bet* presented to the Lord.

"And the Lord said to cosmic *Bet*, 'Most assuredly, *Bet*, for your intrinsic characteristic is the perfect model by which the creative process and the world can achieve their *tikkun*.'" This already has been hinted at by the psalmist when he declares, "For I have said, *Chesed* shall be built up forever."[494]

"The coded word 'shall be built up' also means 'understanding.' The Lord established cosmic *Bet* as a criterion by which to discriminate between those forces or manifestations that are connected with the Lightforce and those entities or manifestations that are linked to the Dark Lord. When humankind is drawn to the energies of imparting–restriction, cosmic *Bet* furnishes the Light of Blessing, as indicated in the verse: 'And test me now herewith,' said the Lord of hosts, 'if I will not open for you the windows of Heaven, and pour out blessing that there not be room enough to receive it.'"[495]

In our discussion of time, we noted that tomorrow is, of necessity, already included in yesterday. The principle of cause and effect, the seed already being included in the tree, the double helix of DNA—all these examples in their separate ways clearly demonstrate that the future depends on our ability to "see" it. The future is here and now, but the illusory one percent of our minds causes us to be blind to its presence. This one percent represents lack and incompleteness. While this one percent was essential to the cosmic process in that it imparted to us the free will necessary to relieve Bread of Shame, the resulting illusion also made us unable to see things as they truly are.

Bet's blessings fill the entire universal space of those individuals who are linked to her. The blessing-force of cosmic *Bet*, implanted in and pervading the cosmos, allows those who are connected with her energy to transcend the illusory defects within the creative process. This situation can be likened to an electric bulb. The moment that the light is switched on, the poles and filament seem to disappear. In deepest truth, the functions of the opposing polarities and the filament are illusory because the light that supposedly appeared only after the switch was flipped was already there in the first place. Light is everywhere: in the middle of a mountain, in the depths of the sea. Our obvious perception of the Light appears as a result of our intervention, the physical effort of flipping the switch, which then disappears, just as the physical components of the bulb seem to disappear. The physical effort involved is only one of the many examples of the illusory creative process that makes us believe that we are the creators of things around us. This illusion arose due to the original restriction that came about because of Bread of Shame.

The internal energy-force of egocentricity, which is just another of our illusions, was established for the express purpose of helping us relieve Bread of Shame.[496] Imparting (positive pole of a light bulb) and restriction (filament of a light bulb) is all the mechanism needed to restore the original all-embracing unified Light, removing it from its illusory realm of darkness. This is why the prophet Malachi declares that if we put the Lord (the all-embracing unified whole) to the test by establishing the two energy-intelligences of imparting and restricting, we shall see that the Lord will "open for you the windows of Heaven, and pour out blessing, that there not be room enough to receive it." Just as an electrician will repeatedly test a circuit or a mechanism to verify that it is functional, so does the prophet Malachi describe a process of testing to determine whether there is any lack or indecision in the blessing-energy of cosmic *Bet*. Any such tests will reveal that blessing is constant and forever present, much like an electric current that is drawn into a home. The electricity is always there, but there are times when an internal connection with the electricity, such as the turning on of a light switch, has not been made. Where a connection has been established with the blessing-energy of cosmic *Bet*, it is said to have been "built up forever" and that an unfailing "continuity" exists. Consequently, there is never "room enough to receive it," for it constantly fills everything it touches, with no room for more.

Now that we understand the true nature of blessing—uninterrupted, continuous fullness—we are in a position to understand that the internal essence of change is defect, lack, deficiency, and incompleteness. Throughout history, various leaders have recommended change as the cure-all for better times ahead. We have been told that to understand our multifaceted human existence, we must shift our perspective from the notion of static social structures to a perception of dynamic patterns of change. Transformation is thought to be an essential step in the development of civilization. Proof of the necessity of transformation, we are told, is the fact that all civilizations have undergone cyclical processes of genesis, growth, breakdown, and disintegration. Arnold Toynbee, in his analysis of the genesis of civilization, concluded that a transition from a static condition to a dynamic one has always existed in the developmental process of every culture.[497] Toynbee described the basic pattern of the genesis of civilizations as "challenge and response." For Toynbee, an effective response to the challenges generated by the genesis process necessitated change and new creative adjustments.

The kabbalist observes in this kind of reasoning the pervasive force of the Dark Lord. You cannot prove the validity of a theory by pointing out how well it explains a given effect. This line of logic may be compared to the fallacious conclusion that Jerusalem is the Holy City because the Holy Temple was located there. The kabbalist asks, "Why was the Temple there in the first place?" The kabbalist insists that one must seek the primal causes of things, rather than getting hung up on the effects.

Jerusalem is the Holy City because it is *whole*. The *Zohar* states that Jerusalem is the energy center of the universe. Therefore, as a natural consequence, the Holy Temple was located in Jerusalem. When change took place in Jerusalem, such as the destruction of the Temple, the presence of the Dark Lord became evident. Any changes that occurred in Jerusalem were reflected in the effect—the Temple—and not in the holiness of the city of Jerusalem itself. The kabbalist considers these changes part of the illusory reality that can only exist when negative human activity severs mankind's connection with the blessing-energy of cosmic *Bet*, and the Dark Lord becomes the ruler of the terrestrial empire. There is no city in the world where so much bloodshed has taken place. Nonetheless, the beauty, importance, and mystique of Jerusalem continue uninterrupted despite the many nations that have temporarily imposed their will upon her.

Why have so many nations considered Jerusalem to be the grand prize in the quest for power? The names and faces have undergone illusory changes, but the real participants followed a precise pattern: They were hungry for power. Jerusalem, the energy center of the world, is the station for the blessing-energy of cosmic *Bet*. This energy caused the convergence of all of Earth's powerful empires on Jerusalem's doorstep. These nations represented an aspect of *klipot* because their primary concern was to connect with the blessing- energy of cosmic *Bet* to repair their defects and fulfill their wanton, covetous needs. Their Desire to Receive remained insatiable and unfulfilled.

There was no evolutionary process involved in these "changing" situations. They followed the pattern of the Dark Lord, whose primary concern was to instill the energy-intelligence of Desire to Receive for Oneself Alone. The result was constant, static uniformity, which ultimately caused the intruding forces to be severed from the blessing-energy of cosmic *Bet* and thus from Jerusalem.

The pattern described seems to fit our current situation quite well. We constantly hear theories concerning our evolutionary process in this new age. And yet King Solomon spoke the truth when he declared, "The thing that has been, it is that which shall be; and that which is done is that which shall be done: and there is no new thing under the sun."[498] The prophet Malachi also spoke eloquently when he said that blessing is ever-present and all-pervading, leaving no room to receive more.

Now we can achieve a deeper understanding of the familiar phrase: "when the novelty wears off." How often have we gone out to purchase some expensive thing, only to find that as soon as it arrives, or shortly thereafter, we no longer want it so badly? Why does this happen? The reason, according to the *Zohar*,[499] is a lack of communication with cosmic *Bet*, the energy-intelligence of blessing. When we are connected with the blessing-energy of cosmic *Bet*, meaning when human activity is in a frame of sharing–restriction, the negative energy of the Dark Lord does not prevail.

Doubt, uncertainty, and lack of fulfillment are the trademarks of the Satan that sometimes controls humankind despite the illusory, egocentric response that it simply "changed its mind." However, when human activity manifests sharing and restriction, the resulting assault by positive energy-intelligence is too much for the Dark Lord to deal with. In reality, there are no changes, no deficiencies, no doubts or illusions. When connected with blessing energy of cosmic *Bet*, man achieves a beautiful clarity. This is the true encompassing energy-intelligence of cosmic *Bet*, the ultimate blessing.

When we look around us, we must conclude that people's lives are largely immersed in illusion, fear of want, unfulfilled dreams—in a word, change. What, then, does reality consist of? Blessing. No change. This is the idea the prophet Malachi describes when he says, "For I am the Lord, I change not; therefore, you sons of Jacob are not consumed."[500]

How much time and effort is "consumed" by the unknowable? The despair and frustration generated by our attachment to illusion are incalculable. Are these negative feelings necessary components of existence? The words of prophet Malachi are quite clear: So long as a cosmic connection exists between ourselves and the Lightforce (the Lord), there are no changes because changes exist only in the world of illusion, the domain of the Dark Lord.

Let us now explore the Satan's inability to penetrate the security shield of cosmic *Bet*. From the *Zohar* has emerged the unique quality of cosmic *Bet*. She was an instrument by which uncertainty, doubts, and illusions would not become manifest so long as mankind maintained its connection with the Lightforce. Selfish activity creates a program of uncertainty in which even the most seemingly flawless plans become subject to indecision. However, when the inherent nature of an individual consists of the energy-intelligence of imparting–restriction, he or she can then access cosmic *Bet* as the program for his or her daily existence. When this occurs, everything improves, even beyond the best laid plans of the individual. The blessing-energy of cosmic *Bet* removes all the rough edges and replaces any doubts or uncertainty with a blessing.

Let us reach into the *Zohar* for a deeper interpretation of the energy-intelligence of cosmic *Bet* and the true nature of blessing:

> *Rav Chiya then began to discourse on the text: "When you have eaten and are satisfied, then you shall bless the Lord."*[501] *Said he, "Should a man then bless the Lord only after he has filled his belly? No, even if one eats but a morsel and 'considers' (desires, meditates) it as a complete meal, this is referred to as eating to satisfaction: as it is written: 'You open Your Hand and satisfy the "desire" of every living thing.'*[502] *The verse*

> *does not state: 'You satisfy with a substantial*
> *meal,' so it is not the quality of food but the*
> *intention of it that 'satisfies;' therefore it is*
> *necessary that at all times when we eat, we*
> *should bless the Lord, so as to provide joy to the*
> *cosmos.*[503]

When we read the Bible literally, we learn that we are to thank the Lord for our daily bread. The verse, however, states very clearly that "when you have eaten and are satisfied, then you shall bless the Lord." As a fundamentalist, I am hard-pressed to understand the verse any other way than how it is stated, namely, to bless the Lord after we have eaten. Furthermore, although we have eaten, we still are not required to bless the Lord until we are satisfied, which is a far cry from the traditional Judeo-Christian interpretation of the verse.

In the *Zohar*, Rav Chiya decodes this very complex and abstruse verse by hinting at another meaning for the word "satisfaction." We know all too well that for many, a morsel of food may be satisfying, and yet for others, there is seldom enough food on the table. Therefore, Rav Chiya concludes that the notion of "satisfaction" is completely bound up with the desire of the individual and not with the amount the Lord has provided. Rav Chiya's interpretation makes it clear that the psalmist paves the way in providing an insight into the world of reality by stating that satisfaction is not in the physical, illusory realm of the actual food, its quantity or quality, but rather, in the manifestation of the blessing—cosmic *Bet*.

Rav Chiya comes to this conclusion because the psalmist correlates that satisfaction depends upon the desire of the individual. In addition, the idea of blessing the Lord might not sit well with many readers. Does the Lord really need our blessing? The reverse might seem more appropriate. One answer to this dilemma is to replace the translation of the Hebrew word *Baruch* (Bless) with "Thank You, Lord." Corruptions of this nature are to be found throughout the Scriptures, where the written word, as conventionally translated, does not provide a satisfactory explanation. But as kabbalists, we know that the Bible represents the cosmic code of our universe, and the *Zohar* struggles successfully to decipher the code: "The blessings that man blesses the Lord with are the metaphysical connections by which Light is drawn from the wellspring of Life—cosmic *Binah* to the Holy Name (Desire to Receive)[504] of the Lord—and to pour down from the Supernal oil and from there to be drawn to the entire universe."[505]

As we've seen, the *Zohar* presents a striking interpretation of the concept of blessings. The word "blessing" is simply another biblical code, and in this case, the *Zohar* states that "blessing" is the code for the connection with cosmic *Bet*. When desire or intent is connected to the framework of restrictive sharing, the individual is thus connected with the blessing-energy of cosmic *Bet*—a state wherein a morsel of food (or a morsel of anything else) imparts complete satisfaction. One feels no hunger when connected with the blessing-energy of cosmic *Bet*, no matter how little he has eaten, nor will he feel indigestion because of overeating. The blessing-energy of cosmic *Bet* leaves no room for lack, defeat, deficiency, or uncertainty. Cosmic *Bet* leaves nothing to chance; she leaves no stone unturned. Never does she find it necessary to change her plans. Everything cosmic *Bet* does is right from the outset.

At the same time, we as individuals have been provided with the free will to either experience the hell of uncertainty, thus becoming an associate of the Dark Lord, or to connect with the Lightforce of the Lord. No matter how we spend our time in this world, the uncertainty that arises from free will shall always be an intrinsic part of our existence. We shall often be plagued by a desire to turn the clock back, to undo what we have done. This is what the *Zohar* alludes to when it says: "The Dark Lord was 'sterile' and had no capacity to yield the fruits of continuity."[506] The Dark Lord is sterile inasmuch as the Desire to Receive for Oneself Alone does not contain an energy-intelligence of continuity. When desire turns into an energy-intelligence of the Desire to Receive for Oneself Alone, the result is an abrupt halt to the flow of the Lightforce of the Lord.

The Desire to Receive for Oneself Alone, represented by the negative pole of an electric bulb, is a classic example of the Dark Lord in action. The filament represents the energy-intelligence of restriction. When the filament of the bulb becomes nonfunctional, we observe a black spot around the area where the malfunction became manifest. Observe further that for a split second after a filament burns out, one can still see the evidence of electric current passing through one pole of the circuit. However, because the connection has been severed, there is no continuity and energy no longer flows. The Dark Lord, in the form of darkness, can now penetrate the bulb's security shield, and the bulb ceases to give the appearance of providing light. Discontinuity, the intrinsic characteristic of the Dark Lord, has set in, taking over the bulb's mini-universe. The broken filament, interruption, and uncertainty are all trademarks of the Satan. When these hold sway, the light must cease to be.

Another example we can use to advance our understanding of the Dark Lord is the mundane act of turning a television or radio on or off. The on position represents the activation of the energy-intelligence of imparting–restriction. When the knob is turned to off, the aspect of imparting–restriction ceases to be operational. However, the broadcasting station is not affected in the least by the knob being turned to an off position. It continues to broadcast, regardless of whether we are receiving. So it is with cosmic *Bet*. When our connection to the blessing-energy of cosmic *Bet* is severed, the Satan can invade and nourish himself by siphoning the Lightforce-energy that has been redirected as a result of the interrupted circuit.

Now, uncertainty becomes a factor we must contend with. When the television has been turned off, the conclusion of the program that was on remains uncertain. The viewer can only guess at the ending. However, one viewer's uncertainty does not dictate uncertainty for others. Those who maintain their connection to the broadcast will suffer no uncertainty, but for the viewer whose connection was severed, there exists a temporary illusion that broadcasting has ceased.

Were the viewer capable of maintaining a link with the broadcast—with the blessing-energy of cosmic *Bet*—the uncertainty most people commonly experience would not exist. Unfortunately, most people do not maintain a constant energy-intelligence of imparting–restriction, and consequently, their connection with the blessing-energy of cosmic *Bet* is severed. Those of us who maintain a constant link with cosmic *Bet* establish a connection with all that is being "broadcast" throughout the entire universe. The bond with cosmic *Bet* provides the assurance that the Dark Lord cannot penetrate the security shield established by the self-administered energy-intelligence of imparting–restriction. The illusory ideas of uncertainty, lack, defect, and discontinuity have no place in the world of reality. The realm of imperfection is the exclusive domain of scientists and others who might like to join the uncertainty club.

This is precisely the concept presented by the *Zohar*: "Rav Shimon spoke on the verse: 'And You, Lord, be not far from me; my strength, haste You to help me.'"[507] Said he [Rav Shimon], "The two invocations 'You' and 'Lord' represent *Malchut* and *Tiferet* respectively, or the two worlds, one of illusion (*Malchut*) and one of reality (*Tiferet*). The psalmist prayed that they become united and not separate from each other. For when the one separates from the other, all Light is darkened and removed from the world. Our universe receives its nourishment, positive energy from *Malchut*, but when *Malchut*,[508] which is the Lower Light, does not receive its strength from *Tiferet*, she has nothing to offer this universe.

"For this reason, the Temple was destroyed in the time of Jeremiah; humankind caused a severance in the link between *Tiferet* and *Malchut*." The awesome power of cosmic *Bet* had been abandoned by human negative activity.

The preceding section of the *Zohar* shows us the dual activity that appears to be present as potential in all energy-intelligent activity. This is precisely the mystery within the subatomic world that drives physicists to distraction. The *Zohar* drives home the point that if negative human activity prevails, then an illusory severance between energy (*Tiferet*) and matter (*Malchut*) seems to take place. The reason behind this cosmic activity—which becomes manifest at all levels from the cosmos down to the realm of subatomic activity—is that humankind has severed the cosmic bond with the blessing-energy of cosmic *Bet*, resulting in a takeover by Satan. Thus, the Dark Lord sets in motion the illusory disintegration of matter, uncertainty, and discontinuity. Still, let us remember that a parallel universe exists for those who maintain their connection with the blessing-energy of cosmic *Bet*. For these fortunate ones, continuity, fulfillment, and the joy of certainty never cease.

The *Zohar* provides an intuitively pleasing picture of our universe, an image that points toward an ultimate simplicity. In addition to providing an amazing description of our environment, Rav Shimon illustrates the idea of two parallel universes bound up and linked to cosmic *Bet* when he discusses the ten martyrs:

"Lord, how did Your children, the foundation of the universe, upon whom the world depended (and I became crowned through their good deeds) suffer such a degrading death at the hands of the Dark Lord. How were their spirits so debased? Despite the fact that when the ten sages were led to their inhuman deaths and their holy bodies were 'substituted' with bodies of the Dark forces, nevertheless, your Holy Name was desecrated. Those who witnessed the slaying were under the impression that the holy bodies of these sages were being tormented."[510]

The martyred sages experienced the peculiar sensation of having "two" bodies: the illusory body, the physical one subject to pain, which belonged to the realm of the Dark Lord; and the holy one, connected with the blessing energy of cosmic *Bet*. Strange as this phenomenon might seem to many readers of this book, I state here that the *Zohar* is replete with many similar illustrations of two parallel universes: the world of reality and the realm of illusion.

Prior to the advent of quantum physics, "illusion" meant an incorrect perception of external stimuli, such as the mirages seen in a desert or the moment when a straight spoon appears to bend when submerged in a glass of water. Today, however, with uncertainty established as an integral component of reality itself, illusion no longer necessarily implies distorted transmission or interpretation of stimuli. What is perceived by the individual may, in fact, be considered reality, and what most of us perceive as reality may instead result from the false impressions conveyed by the illusory world.

It is meaningless to argue about which view is correct because nature's elemental properties defy objective evaluation, as quantum physics has taught us with discoveries like Heisenberg's Uncertainty Principle. The first conclusion we tend to draw from uncertainty is that there is a "real" domain beyond our perceptions, but that man, with his present limitations, is not capable of entering this domain. Some will therefore conclude that quantum uncertainty and the consequent failure of the physical law of cause and effect provide the solution to the problem of free will. The atheist, on the other hand, will now have found justification for his contention that chance rules the universe.

The *Zohar* and cosmic *Bet* seem to be the only way out of this dilemma. Cosmic *Bet* is the link to certainty, the realm beyond the clutches of the Dark Lord and all that he represents. The profound implications and benefits of the blessing-energy of cosmic *Bet* are too numerous to allow us to even begin to conceptualize what effect the new realm of reality might have on our overall mental and physical well-being. Indeed, the blessing-energy of cosmic *Bet* connects us with the idea of holistic medicine, which takes into account all aspects of a patient's life in the final diagnosis and treatment. Cosmic *Bet* is the link to completeness.

Another fine example of the way the Dark Lord can manipulate our sense of reality is found in the *Zohar*'s description of the biblical account of the golden calf, an idol that Aaron made at the behest of the Israelites waiting for Moses at the foot of Mount Sinai.[511] When the Lord learned of this, He told Moses of the apostasy of the Israelites, whom he then proposed to destroy. Moses, while carrying the Tablets of the Covenant down from Mount Sinai, saw the people dancing around the golden calf. In great anger, Moses smashed the Tablets, and then melted down the calf idol, pulverized the precious metal, and scattered the powdered gold into the drinking water of the Israelites, thus ensuring the people would ingest it.

However, the *Zohar* departs from the original narrative in one significant way: The *Zohar* claims that the Tablets were never shattered. Rav Akiva said to his students, "Do not compare pure marble to other stone which contains the energy-intelligences of life and death. We know this from the Scriptures: 'A wise man's heart is at his right hand, but a fool's heart at his left.'[512] The original Tablets were made from pure marble material, and therefore, no separation or discontinuity was present. It was only an illusion to consider the Tablets as having been broken."[513]

How did this illusion come to pass? Was Moses some kind of magician? Magicians do their work by reawakening the realm of true reality that lies dormant in all of us. As children, we were fascinated by a leaf changing colors. It seemed to us that we were witnessing a miracle until we were taught that we were merely observing the process where the leaf loses its chlorophyll, causing it to change color and die. We were thus taught not to be fascinated by such things. But perhaps it is time once again to take a second look at many things we learned not to question.

The Bible states clearly that when Moses descended from Mount Sinai, "He cast the Tablets out of his hands, and he broke them beneath the mount."[514] Thus, there seems to be an apparent contradiction between the *Zohar*'s interpretation of the shattering of the Tablets and what the Bible says. Once again, the *Zohar*[515] brings to our attention the two realities that became established when Adam sinned. The Tree of Life represented the true reality, whereas the Tree of Knowledge of Good and Evil brought the world of illusion into existence. When Adam became severed from cosmic *Bet*, the realm of reality became concealed by a veil of illusion, and thus the world of uncertainty was born.

In essence, the Tablets were never shattered; but were placed in the Ark of the Covenant. When the High Priest accessed the Tablets, he connected with the cosmos, drawing awesome power from it. For those who participated in the sin of the golden calf, the veil of illusion—the energy-intelligence of uncertainty—became their realm of reality.

Another example of the illusory world is the mistaken belief that the Holy Temple in Jerusalem was set afire by the Romans and destroyed. Not so, claims the *Zohar*. If ever the intelligent energy of illusion became manifested in its totality, such an event is revealed in the *Zohar*'s contention that the Holy Temple was never destroyed. The historical account of the Roman conquest of the Holy Land, the destruction of the Temple, the deportation into slavery of more than a hundred thousand Israelites, and the slaughter of hundreds of thousands more remains a vivid testimonial to this day to the destruction of the Temple. Anyone who visits the Temple Mount today will not visually observe a Temple there. However, with uncertainty becoming an integral part of our understanding of reality, we must admit that one cannot disprove the *Zohar* on this point that according to it, the essence of the Temple is still there; only the illusion was destroyed.

Despite their conclusion that everything in our observable universe is uncertain, scientists still pursue a theory of grand unification. Why? Because something tells them that beyond quantum uncertainty lies a deeper, true metaphysical reality. The uncertainty principle applies to a single measurement as well as to a statistical average. To achieve the realm of deterministic reality, one must raise one's level of consciousness. This requires a connection with the blessing-energy of cosmic *Bet*, which transcends the illusory frame of reference.

Quantum discoveries did not, as physicists believe, prove that the universe is indeterminate. Quantum merely indicates that our conscious mind cannot grasp all of reality at once. A new vision of reality is now necessary. The *Zohar* contends that the answer to this and to all physical dilemmas lies in the transcendence of the conscious mind, which itself is a product of the illusionary reality. There is no randomness once our consciousness has been elevated. Unfortunately, this is not the path science has chosen to follow.

The *Zohar* continues in its explanation why the Temple was never destroyed: "Over those stones of the former foundations other nations could prevail, inasmuch as they lacked a higher consciousness of energy-intelligence…Heaven forbid, thinking for a moment that other nations ruled over the foundations of Zion and Jerusalem. They did not torch them, nor were they burnt. They became concealed by the Lord, and not even one stone was lost. However, when the Light of reality returns, the only people to observe it will be those whose eyes have been raised to a higher consciousness. For the others, the illusion will reign."[516]

This is precisely the idea the prophet Malachi presented when he declared, "You shall return and discern between the righteous and the wicked, between him that serves the Lord and him that serves Him not."[517] In the absence of a better translation, I have accepted the conventional version of this profound, penetrating verse. However, make no mistake that the idea of serving the Lord is to be understood within a framework of connection with the blessing- energy of cosmic *Bet*. When a link between us and cosmic *Bet* has been established, there is certainty. The idea of uncertainty is initiated solely by the Desire to Receive for Oneself Alone.

"The Creator did not reveal the Light of Blessing to bring about the perfection and correction of the world. Instead, the Light of

Blessing serves only as a good beginning, which is essential in bringing about the final and complete correction."[518]

Questions about the nature of reality can be subtle indeed. It can be extraordinarily difficult to conclude just what is real and what is illusion. The truths contained in the *Zohar* broach the possibility that we will one day make sense out of our existence. All that is required is a connection to cosmic *Bet*. This was the blessing cosmic *Bet* could provide for all mankind. The connection had yet to be established, but cosmic *Bet* was a good starting point for achieving this blessing.

Blessing;
Stability in the Universe;
Knowledge as Energy;
Deficiency of Language;
Uncertainty Principle;
Macro and Micro Worlds;
Direction of Meditations;
Quantum Theory;
Human Suffering;
World of Illusion;
Negative Polarities;
Circle and Waves of Energy;
Seed and the New Fruit;
Resistance and Intensity;
Making of a Boom;
Positive Activity;
Three Upper *Sefirot*

CHAPTER 22

THE LETTER
ALEF

IT IS NOT YOUR PART TO
FINISH THE TASK;
YET NEITHER ARE YOU FREE
TO DESIST FROM IT.

—RABBI TARFON, *PIRKEI AVOT* (ETHICS OF THE FATHERS)

C osmic *Bet* with the energy-intelligence of blessing was the ideal channel for the network of universal communication. Only she could prevent humanity from becoming eternally disconnected from the Lightforce. Now that cosmic *Bet* had been established as the creative channel, humanity would never have to succumb to total robotic consciousness, and the Dark Lord would be prevented from dominating the universe by removing all vestiges of free will from the human psyche.

With the matter of universal Creation and stability well in hand, cosmic *Alef* had no cause to come forward to enter a plea to become a suitable channel for Creation. Nor did she have this desire, for she understood and was satisfied with her role in the cosmic scheme. She was to be a silent partner in the realm of thought-communication. It would be her function to link the other letter-energies, bridging the illusion of space between them.

From the onset, the Lord recognized the purpose for each letter's participation in the creative process. The letters— whether they would be chosen as the channel for the creative process or not—were ultimately a part of the celestial mechanism for communication with the supreme energy of Cosmic Intelligence. Obviously, then, the Lord was aware of *Alef*'s actual place in the universal plan. And yet He still saw fit to ask *Alef* why she did not come forward.

Each letter, including cosmic *Alef*, was necessary for the ultimate connection with Cosmic Intelligence. The Lord had already chosen to use cosmic *Bet* as the channel for His Creation. Why then did the Lord call upon *Alef* to come forward?[519]

Cosmic *Bet* established control over the process of the formation of the universe because her *berachah* (blessing-intelligence) guided the creative process, allowing life on Earth to inherit a favorable setting for its development. The cosmic intelligence operating outside the material realm, guided by *Bet*'s *berachah*-intelligence, would be sufficient for maintaining the proper balance and symmetry between positive and negative polarities, good and evil. This cosmological phenomenon, which is an essential element of universal balance, was Earth's guarantee that Satan would never assume dominion over this vast cosmos.

Cosmic *Alef* is the intrinsic thought-intelligence of air, which establishes the steady-state framework of our universe.[520] The fact is that in our universe, nothing really changes. The original Big Bang was only the most intense of an infinite number of miniature big bangs that were to follow, after which the universe would be returned to a unified, steady state of activity by cosmic *Alef*. Whether we are speaking of explosions involving high-density matter, low-density gases, or no-density metaphysical phenomena, the basic underlying principle is the same. In all cases, the unifying steady-state factor is *Alef*.

Let us examine the process of picking up a pebble and tossing it into the sea. Because the process first takes place in the mind of the person who is going to throw the pebble, the thrower's desire must be included as an integral part of the process. The outcome is determined in the mind of the thrower before he or she tosses the pebble. The result, with a few minor variations, is predetermined, established in the mind and carried out by the law of cause and effect. The stone plays a fairly insignificant role in the process. What, then, causes the splash and subsequent ripples? The kabbalist says the causal factor is the thrower's mind.

As for the circles of waves that form around the pebble when it strikes the water, we might expect the largest ripple to occur at the point of impact because that is where the pebble's kinetic energy is concentrated. Yet we encounter the largest waves farther away from the pebble, not closer to it. The larger, outer waves represent the thought-intelligence of the thrower. The thought-intelligence at its starting point—when the individual decided to throw the rock—carries with it a great deal more energy than those stages subsequent to the initial thought.

A similar situation exists with regard to the making of a bomb. The bomb explodes first in the mind of its inventor, who then proceeds to work on ideas for how that explosion might be produced on a physical level. The explosion then progresses through a series of thought processes within the minds of those who are producing and developing the bomb, gaining power as it nears completion. Finally, perhaps years later, when the bomb is detonated, the explosion (restriction) reveals the thought-energy-intelligence that existed first in the inventor's mind.

Before detonation, the bomb is in a steady state of thought-intelligence. It is only when the bomb strikes a target that the thought-intelligence is revealed on the corporeal level. At all levels of interaction between the thought-energy and its Vessel (the bomb), the same dynamic "big bang" interplay recurs, the only significant difference being in the nature of the materials, the density of the matter, and the heat of the interaction. Whether we are discussing a pebble striking water, an electrical current passing through a filament, or a nuclear reaction, all are determined according to the same fundamental principles governing the operation of thought-energy.

A growing tree undergoes the process of restriction at each stage of its development. Before the trunk of a tree appears, the expanding root encounters restriction, which in turn encourages other growth. This process repeats continually from the moment of planting until the final fruition. Indeed, expansion continues even after that. The seed of the new fruit undergoes another big bang when it falls to the ground or someone inserts it into Mother Earth.

The universe has encountered the thought-intelligence of restriction many times during the course of its expansion from the original Big Bang, the impact point that serves as a target for the next stage of universal evolution. The uniformity and predictability of this universal process are caused by the energy-intelligence of *Alef.*

Where is the drive for sustained activity coming from? Why, for instance, do parents desire children when they know very well the pain and heartache this activity will cause? The answer lies in the original Thought of Creation, which was the Lord's Desire to Impart.[521] This ongoing activity manifests itself at all levels of existence within the universe. Consider, for a moment, the fact that the basic building blocks of both thinking and non-thinking entities are the same kinds of atoms. What, then, distinguishes one from the other? We can answer this question by citing the different arrangements of the atoms, but the kabbalist inquires further, asking why should these different arrangements exist, and what is the cause for their dissimilarity? The true difference between animate and inanimate beings lies in the level of intensity of their Desire to Receive.[522] The greater the desire, the greater the necessity for restriction. Thus, a balance is maintained between sharing and receiving, with the consciousness of cosmic *Alef* governing this balance.

Obviously, cosmic *Alef* fills a crucial need in the maintenance of the universe. Why, then, did she fail to plead her case before the Lord? Also, as we shall see, when the Lord addressed cosmic *Alef*, He repeated her name twice. Why did He do this?

All of the twenty-two letters provide links to the Lightforce, but *Alef*'s link is very different from the rest. The other letters' channels provide access to the Lightforce only when the initiative comes from the Vessels—the letter-energies themselves. Cosmic *Alef*, however, cannot initiate *Mayin Nukvin* (the Vessel's energy-intelligence of Desire to Receive). *Alef*'s energy is derived solely from the Lightforce.

This unique characteristic of cosmic *Alef* was revealed by the *Zohar*'s interpretation of the verse in the book of the prophet Amos:[523] "The Virgin of Israel [*Shechinah*] is fallen; she shall no more rise."[524] The *Zohar* declares that the *Shechinah* cannot rise from exile through her own effort. She shall be redeemed only by the Lord Himself. According to the *Zohar*,[525] the *Shechinah* will undergo two periods of redemption: "See now. In all the other exiles of Israel, a term was set, at the end of which Israel returned to the Lord, and the Virgin of Israel came back to her place. But this last and present exile is not so, for she shall not return as on previous occasions. The verse in Amos indicates this also when it says: 'The Virgin of Israel is fallen, she shall rise no more.' Note that it is not written: 'I shall not raise her anymore.'"

The *Zohar* describes a king who was angry with his queen and banished her from his palace for a very long time, until finally he could no longer bear her absence:

> Said the king, "This time is not like the other times when she came back to me. This time, I shall go with all my followers to find her." When he came to her, he found her lying in the dust. Seeing her thus humiliated and yearning once more for her, the king took her by the hand, raised her up, brought her back to his palace, and swore to her that he would never part from her again. So it is written: "In that day, I will raise up the Tabernacle of David that is fallen,"[526] the Tabernacle of David being identical with the Virgin of Israel.

Cosmic *Alef* will never initiate the *Mayin Nukvin*. While the other letters express the *Mayin Nukvin*, cosmic *Alef* will act only as the channel for the Lightforce of the Lord to return *Mayin Duchrin* to the infinite circular condition from which it first emerged.

Therefore, cosmic *Alef* did not enter her plea to be the channel for Creation. Yet the Lord called her name twice: "*Alef. Alef . . . .*" The Lord called her name once for the purpose of transmitting the Lightforce through *Alef*, when human positive activity prevailed. Then the Lord called her name a second time to establish the channel for the Final Redemption, telling her, "You, *Alef*, shall be considered the Head of all the letters, for you represent the Upper Three *Sefirot* [*Keter*, *Chochmah*, and

Binah]. The great unifying energy of the Lightforce, the Head, will be revealed only through you."[527]

The energy infused within cosmic *Bet* embraces the force of Genesis I, the Light of Mercy, which includes only the Lower Seven *Sefirot* of *Chesed, Gevurah, Tiferet, Netzach, Hod, Yesod,* and *Malchut* that govern this World of Action. Consequently, the biblical code presented in Genesis I is limited to the seven days of physical Creation. Cosmic *Bet* is devoid of the Lightforce of the Upper Three *Sefirot*, also called the Head. Cosmic *Bet*'s domain is the world of the created illusion. It is cosmic *Alef* who provides the vital link with the true Infinite Reality, the connection with the Head *Sefirot*, also known as the Light of Wisdom.

Bet's Light of Mercy is not dependent upon mankind's activity. When mankind expresses and makes manifest negative thought-energy-intelligence, the power of cosmic *Bet* does not change. Cosmic *Bet* maintains the Light but has no power to alter its course or dimension. Like a thrown pebble after it leaves the hand of the thrower, cosmic *Bet*'s Light of Mercy is no longer governed by the factor that caused it. Having left its source, it is now controlled by the law of cause and effect. The *Mochin* (the Lightforce of the Head), on the other hand, channeled by cosmic *Alef*, depends entirely upon human activity. If man is evil, the *Mochin* of *Alef* is rescinded. But when positive human thought-activity prevails in the universe, then *Alef* establishes the grand unification Lightforce of the Lord. Complete and eternal unification will take place at the time of the Final Redemption and Correction. This, too, will manifest with the aid of the "silent partner," cosmic *Alef*.

CITATIONS

Introduction

1. Gate of Divine Inspiration, Writings of the Ari, Vol. 12, KCI, P. 10.
2. Implications of Meta-Physics for Psycho-Energetic Systems, Jack Sarfatti.
3. Gate of Divine Inspiration, Writings of the Ari, Vol. 12, KCI, P. 5.
4. The Physicists Conception of Nature, P. 224.
5. Ideas and Opinions, Albert Einstein.

Chapter 1

6. Superspace and the Nature of Quantum Geometrodynamics, J.A. Wheeler.
7. Smithsonian and National Academy of Science, J.A. Wheeler.
8. Zohar, Naso, Idra Raba 9:65.
9. Jeremiah 31:33.
10. The Mysterious Universe, Sir James Jeans.
11. Zohar, Prologue 6:22.
12. Exodus 34:19.
13. Zohar, Beresheet A 22:254

Chapter 2

14. Psalms 33:6.
15. Exodus 35:10.
16. Talmud Bavli, Tractate Berachot P. 55b
17. Exodus 35:31.
18. Genesis 4:1
19. Genesis 5:3.
20. Song of Songs Rabbah 5:14.
21. Talmud Bavli, Tractate Avoda Zara, P. 182.
22. Genesis 4:14.

23. Genesis 17:4-16.
24. Genesis 15:5.
25. Zohar, Pinchas 11:65
26. Zohar, Ki Tetze 3:32
27. Psalms 19:2-5.
28. Exodus 2:11-12.
29. Zohar, Shemot 23:207
30. Zohar, Va'era 2:25-28.
31. Malachi 3:10.

Chapter 3

32. Zohar, Beresheet B 44:181
33. Talmud Bavli, Tractate Gittin, P. 68a-b.
34. Numbers 24:17.
35. II Samuel 8:2
36. Zohar, Balak 47: 501
37. Genesis 47:11-27.
38. Zohar, Acharei Mot 45:276
39. Zohar, Vayikra 59:388
40. Numbers 12:6-8.
41. Exodus Rabbah 2:2.
42. Exodus 4:14.
43. Leviticus 11:1, 13:1.
44. Avot DeRabbi Natan, 48.
45. Exodus 32:4.
46. Avot, 1:12.
47. Zohar, Emor 1:2
48. Numbers 16:1-35.
49. Zohar, Pekudei 52:767
50. Genesis 29:20.
51. Zohar, Vayechi 77:783
52. Genesis 37:2.
53. Zohar, Vayeshev 23:253
54. Genesis 37:18.
55. Genesis 39:7-16.
56. Genesis 41:37-40.
57. I Samuel 16:1.
58. Ruth 4:17, 20-22.
59. I Samuel 18:27.
60. I Samuel 18:1-5.
61. I Samuel 18:27.
62. Exodus, 28:30.
63. II Samuel, 2:1.
64. II Samuel 20:1.

65. Genesis Rabbah 88:7.
66. Talmud Bavli, Tractate Sanhedrin, P. 110a.
67. II Samuel Ch. 11.
68. Wheels of a Soul, Rav Berg, KCI, P. 150.
69. Talmud Bavli, Tractate Shabbat, P. 56a.
70. Talmud Bavli, Tractate Pesachim, P. 117a.
71. Talmud Bavli, Tractate Berachot, P. 10a.
72. Talmud Bavli, Tractate Berachot, P. 3b.
73. Zohar, Kedoshim 11:73

Chapter 4

74. Kabbalah for the Layman, Vol. 1, Rav Berg, KCI, Ch. 3.
75. Kabbalah Connection, Rav Berg, KCI, P. 94, 96.
76. Zohar, Toldot 1:3-4
77. Entrance to the Tree of Life, Rav Ashlag, KCI, par. 54-58.
78. Kabbalah for the Layman, Vol. 1, Rav Berg, KCI, Ch. 3
79. Genesis Ch.1.
80. Genesis 14:18.
81. Zohar, Lech Lecha 25:237-240
82. Talmud Eser Sefirot (Ten Luminous Emanations), Vol.1, Rav Ashlag, KCI, P. 75
83. Book of Formation, Cch:1 Mishna 10-ch.38.
84. Book of Formation, Ch.4.
85. Book of Formation, Ch.5.

Chapter 5

86. Talmud Eser Sefirot (Ten Luminous Emanations), Vol.1, Rav Ashlag, KCI, P. 55
87. Psalms 145:1b.

88. Zohar, Ha'azinu,51:210
89. Psalms 103:19
90. Proverbs 5:5.
91. Genesis 3:4-6.
92. Zohar, Beresheet A, 47:442; Tikunei Zohar, par. 77, P. 403
93. Genesis 3:14
94. Zohar, Prologue 6:23
95. Kabbalah for the Layman, Vol. 1, Rav Berg, KCI, P. 81
96. Talmud Bavli, Tractate Shabbat, P. 55a.
97. Talmud Bavli, Tractate Shabbat, P. 55a.

Chapter 6

98. Zohar Hebrew Edition, Prologue, Sulam, par. 24
99. Numbers 8:2
100. Genesis 1:3-4
101. Genesis 1:4
102. Genesis 1:4
103. Zohar, Terumah 78:762
104. Talmud Eser Sefirot (Ten Luminous Emanations), Vol.2, Rav Ashlag, KCI, P. 105-110
105. Genesis 11:31.
106. Genesis 12.1.
107. Genesis 10:8
108. Kabbalah for the Layman, Vol. 1, Rav Berg, KCI, P. 101-104
109. Zohar, Lech Lecha 5:27
110. Kabbalah for the Layman, Vol. 1, Rav Berg, KCI, P. 77-90
111. Talmud Eser Sefirot (Ten Luminous Emanations), Rav Ashlag, KCI, Vol.2, P. 56-57
112. Talmud Eser Sefirot (Ten Luminous Emanations), Vol.1, Rav Ashlag, KCI, P. 121
113. Kabbalah for the Layman, Vol. 1, Rav Berg, KCI, P. 73-75
114. Kabbalah Connection, Rav Berg, KCI, P. 23-27

115. Zohar Hebrew Edition, Lech Lecha, Sulam par. 22
116. Cosmic Tav
117. Tikunei Zohar, Tikun 22 par. 60
118. Numbers 27:21
119. Numbers 28:30.
120. Deuteronomy 33:8
121. Talmud Bavli, Tractate Yoma, P. 73a-b
122. Midrash Psalms 27:2
123. Talmud Bavli, Tractate Yoma, P. 73b
124. Entrance to the Zohar, Rav Ashlag, KCI, P. 54-58
125. Zohar Hebrew Edition, Prologue, Sulam par. 24.
126. Zohar, Vayera 20:280.
127. Zohar Chadash, Shir Hashirim par. 12.
128. Zohar Hebrew Edition, Prologue, Sulam, par. 24.
129. Zohar Hebrew Edition, Prologue, Sulam, par. 24

Chapter 7

130. Zohar, Naso 20:190
131. Zohar Hebrew Edition, Prologue, Sulam, par 25.
132. Zohar, Chayei Sarah 12:72-73
133. Zohar Hebrew Edition, Prologue, Sulam, par 25
134. Talmud Bavli, Tractate Shabbat P. 104a
135. Zohar Hebrew Edition, Prologue, Sulam, par 24
136. Kabbalah for the Layman, Vol. 1, Rav Berg, KCI, P. 88-92
137. Kabbalah Connection, Rav Berg, KCI, P. 104
138. Zohar Hebrew Edition, Prologue, Sulam par. 24
139. Wheels of a Soul, Berg, p.79
140. Ecclesiastes 4:13-14.
141. Zohar, Vayeshev 1:3.

Chapter 8

142. Genesis, 3:8
143. Zohar, Vayikra 1:16
144. Leviticus 1:1
145. Genesis 15
146. Zohar, Pekudei 13:77
147. Zohar, Beresheet B, 30:113

Chapter 9

148. Kabbalah Connection, Rav Berg, KCI, P. 117
149. Genesis 2:9
150. Kabbalah Connection, Rav Berg, KCI, P. 92
151. Talmud Bavli, Tractate Pesachim, P. 112a
152. Talmud Yerushalmi, Tractate Sanhedrin, P. 3a
153. Zohar Chadash, Ki Tavo par. 1
154. Talmud Bavli, Tractate Shavuot, P. 33b
155. Genesis 1:27
156. Genesis 2:21, 22
157. Genesis 3:16-25
158. Zohar, Beresheet B, 64:366
159. Talmud Bavli, Tractate Berachot, P. 17a
160. Talmud Bavli, Tractate Berachot, P. 17b
161. Ecclesiastes 7:26
162. Zohar, Prologue 21:224
163. Zohar, Beresheet B 51: 228-229
164. Genesis Ch.7
165. Genesis 9:20
166. Genesis 7:3
167. Genesis 6:18
168. Proverbs 10:25
169. Genesis 6:9
170. Genesis 6:9
171. Genesis 6:8-9
172. Zohar, Noah 1:10

Chapter 10

173. Zohar Hebrew Edition, Prologue, Sulam, par. 27
174. Genesis 4:7
175. Entrance to the Zohar, Rav Berg, P. 22-27
176. Ecclesiastes 7:14
177. Samuel I 4:22-23.
178. Zohar Hebrew Edition, Prologue, Sulam, par. 27.
179. Zohar, Vayikra 22:136.
180. Zohar, Vayikra 22:135-136
181. Zohar, Shemot 13:75
182. Samuel II 6:17
183. Genesis ch.24
184. Cf. Tractate Megillah, P. 26a
185. Zohar, Trumah 45:486
186. Zohar Hebrew Edition, Prologue, Sulam, par. 27
187. Genesis Ch.11
188. Proverbs 10:25
189. Zohar Chadash, Vayera P. 26
190. Psalms 145:19
191. Kabbalah for the Layman, Vol. 1, Rav Berg, KCI, P. 24
192. Joshua 10:12-13
193. Joshua 6:13-15
194. Exodus, 14:21
195. Rashi, Exodus 14:21
196. Targum Yerushalmi, Exodus, Ch.14:22
197. Exodus 14:19
198. Zohar, Beshalach 13:157
199. Exodus 24:18
200. Exodus 14:19-21
201. Exodus 17:9
202. Exodus 33:11
203. Zohar, Beshalach 33:463
204. Zohar Hebrew Edition, Prologue, Sulam, par. 27
205. Kabbalah for the Layman, Vol. 1, Rav Berg, KCI, P. 77-90
206. Deuteronomy 32:11
207. Kabbalah Connection, Rav Berg, KCI, P. 117-118
208. Genesis 2:9
209. Genesis 2:17
210. Zohar, Beresheet A 46:432
211. Talmud Bavli, Tractate Shabbat, P. 146a
212. Isaiah 25:8
213. Zohar, Prologue, 6:27

Chapter 11

214. Isaiah 58:13
215. Exodus 11:1
216. Physics and Philosophy, W.Heisenberg, P. 177
217. Talmud Eser Sefirot (Ten Luminous Emanations), Vol.1, Rav Ashlag, KCI, P. 52-54
218. Zephaniah 2:3
219. Zohar Hebrew Edition, Prologue, Sulam, par. 27
220. Exodus 28:43

Chapter 12

221. Kabbalah for the Layman, Vol. 1, Rav Berg, KCI, P. 77
222. Zohar, Beresheet B 41: 172
223. Genesis 1:27
224. Song of Songs 2:12
225. Genesis 2:5
226. Genesis 3:17
227. Genesis 4:12
228. Zohar, Beresheet A 23:258
229. Kabbalah for the Layman, Vol. 1, Rav Berg, KCI, P. 97
230. Zohar Hebrew Edition, Prologue, Sulam, par. 28
231. Kabbalah Connection, Rav Berg, KCI, P. 39
232. Zohar, Prologue 12:77
233. Zohar, Vayikra 42:288
234. Zohar, Vayikra 43:297
235. Wheels of a Soul, Rav Berg, KCI, P. 116-129
236. Genesis 1:14-19

237. Zohar, Beresheet A, 7:71-11:129
238. Kabbalah for the Layman, Vol. 1, Rav Berg, KCI, P. 106-108
239. Talmud Eser Sefirot (Ten Luminous Emanations), Vol. 2, Rav Ashlag, KCI, P. 157-158
240. Psalms 145:14
241. Job 38:33
242. Isaiah 26:4
243. Genesis 49:26
244. Deuteronomy 4:32
245. Zohar, Va'era 1:6
246. Zohar Hebrew Edition, Prologue, Sulam, par. 28
247. Kabbalah for the Layman, Vol. 1, Rav Berg, KCI, P. 79
248. Genesis 2:4
249. Genesis 1:2.
250. Zohar, Beresheet A 19:214

Chapter 13

251. Kabbalah Connection, Rav Berg, KCI, P. 112-114
252. Zohar Hebrew Edition, Prologue, Sulam, par. 28
253. Genesis 8:21
254. Exodus 15:11
255. Psalms 33:1
256. Zohar Hebrew Edition, Prologue, Sulam, par. 29
257. Zohar Hebrew Edition, Prologue, Sulam, par. 29

Chapter 14

258. Zohar Prologue, 6:30
259. Psalms 42:9
260. Genesis 1:1-8
261. Zohar Hebrew Edition, Prologue, Sulam, par. 30
262. Zohar Naso 9:65
263. Jeremiah 31:34.
264. Zohar Hebrew Edition, Prologue, Sulam, par. 30

265. Zohar Hebrew Edition, Prologue, Sulam, par. 30
266. Kabbalah Connection, Rav Berg, KCI, 117-118
267. Isaiah 6:1
268. Psalms 81:4
269. Zohar, Emor 32:190
270. Zohar, Beshalach 13:155
271. Kabbalah Connection, Rav Berg, KCI, P. 133-135
272. Exodus 9:1
273. Zohar, Bo 3:38
274. Exodus 20:2
275. Exodus 1:8
276. Zohar, Shemot 13:77
277. Zohar Hebrew Edition, Prologue, Sulam, par. 30
278. Proverbs 12:4
279. Song of Songs 3:11
280. Zohar, Prologue 14:125
281. Exodus 19
282. Zohar, Prologue 14:125
283. Zemirot Israel, Najara, Venice, 1599
284. Sabbath Prayer Book, Evening Prayer
285. Wheels of a Soul, Rav Berg, KCI, P.168-177

Chapter 15

286. Zohar, Prologue 6:31
287. Judges 5:4
288. Psalms 68:8
289. Job 9:5-9:7
290. Gate of Meditations A, Writings of the Ari, Vol. 10, KCI
291. Zohar Hebrew Edition, Prologue, Sulam, par. 31
292. Zohar, Toldot 1:3
293. Talmud Eser Sefirot (Ten Luminous Emanations), Vol. 1, Rav Ashlag, KCI, P. 16, Par. 8-10

294. Zohar, Lech Lecha 4:19
295. Ecclesiastes 7:14
296. Talmud Bavli, Tractate
     Pesachim, P. 6b
297. Kabbalah for the Layman,
     Vol. 1, Rav Berg, KCI, Ch. 9
298. Zohar Hebrew Edition,
     Prologue, Sulam, par. 31
299. Zohar, Prologue 6:31

Chapter 16

300. Wheels of a Soul, Rav Berg,
     KCI, Ch. 17
301. Zohar Hebrew Edition,
     Prologue, Sulam, Par. 1-9
302. Genesis Rabbah, Ch.12
303. Talmud Eser Sefirot (Ten
     Luminous Emanations), Vol.
     1, Rav Ashlag, KCI, P. 14
304. Talmud Eser Sefirot (Ten
     Luminous Emanations), Vol.
     4, P. 218
305. Kabbalah for the Layman,
     Vol. 1, Rav Berg, KCI, Ch. 8
306. Zohar, Mishpatim 3:237
307. Talmud Eser Sefirot (Ten
     Luminous Emanations), Vol.
     1, Rav Ashlag, KCI, P. 31,
     Par. 16
308. Kabbalah for the Layman,
     Vol. 1, Rav Berg, Ch. 6-7
309. Genesis 1:6
310. Kabbalah for the Layman,
     Vol. 1, Rav Berg, KCI, Ch. 9
311. Zohar, Beresheet A, 6:45
312. Genesis 1:10-1:12
313. Zohar, Beresheet A, 8:97
314. Ecclesiastes 7:14
315. The Kabbalah Connection,
     Rav Berg, KCI, Ch. 15
316. Zohar, Chayei Sara 11:70
317. Zohar, Chayei Sara 12:71-72
318. Psalms 139:5
319. Genesis 9:2
320. Genesis 2:17

321. Zohar, Beshalach 17:230-232
322. Zohar, Toldot 17:133
323. Zohar, Terumah 4:20-21
324. Zohar, Shemot 15:140-143
325. Proverbs 10:25
326. Zohar, Acharei Mot 39:233
327. The Kabbalah Connection,
     Rav Berg, KCI, Ch. 10
328. Exodus 2:12
329. Zohar, Terumah 2:8-9, 2:11
330. Talmud Bavli, Tractate
     Pesachim, P. 50a
331. Malachi 3:6
332. Zohar, Prologue 6:32

Chapter 17

333. Zohar, Prologue 6:33
334. Genesis 1:4
335. Isaiah 3:10
336. Zohar, Beresheet A, 32:324
337. Asymptotic Realms of
     Physics, A. H. Guth
338. Wheels of a Soul, Rav Berg,
     KCI, Ch. 1, 2
339. Kabbalah for the Layman,
     Vol. 1, Rav Berg, KCI, Ch. 8
340. Talmud Eser Sefirot (Ten
     Luminous Emanations), Vol.
     1, Rav Ashlag, KCI, P. 56,
     Par. 19
341. Genesis Rabbah, Ch. 9
342. Talmud Bavli, Tractate
     Kiddushin, P. 30b
343. Deuteronomy 28:18.
344. Genesis 1:4
345. Psalms 112:3; Proverbs, 11:31
347. Zohar, Vayeshev 22:230-240
348. Zohar, Vayeshev 21:210-213
349. Genesis Rabbah 99:7
350. Genesis 37:24
351. Genesis Rabbah 91:8
352. Genesis 50:15-50:21
353. Sefer HaLikutim, Writings of
     the Ari, Vol. 18, KCI, Portion
     of Beha'alotcha.

354. Psalms 78:38
355. Isaiah 1:27
356. Deuteronomy Ch. 49
357. Deuteronomy 8:6
358. Talmud Bavli, Tractate Yevamot, P. 99a
359. Mishneh Torah, Rabbi Moshe Ben Maimon (Rambam), Assurei Bia, Ch. 19
360. Isaiah 11:6
361. Isaiah 11:9.
362. The Wisdom of Truth, Rav Ashlag, KCI, 2008.
363. The Wisdom of Truth, Rav Ashlag, KCI, 2008
364. Numbers 11:11-16
365. Zohar, Ki Tisa 10:56-62
366. Zohar, Ki Tisa 11:62-70
367. Sha'ar Hapsukim, Writings of the Ari, Vol. 8, KCI, P. 107
368. Exodus 32:7
369. Exodus 32:12
370. Deuteronomy 34:10
371. Midrash Rabbah, Numbers, Ch. 14
372. Gate of Reincarnation, Writings of the Ari, Vol. 13, KCI, P. 224
373. Exodus 2:2
374. Pri Etz Chaim B, Writings of the Ari, Vol.17, KCI, P. 246-247
375. Mishneh Torah, Rabbi Moshe Ben Maimon (Rambam), Assurei Bia, 14:7
376. Genesis Rabbah Ch. 44
377. Genesis Rabbah P. 24
378. Genesis 46:3
379. Exodus 32:9-10
380. Gate of Reincarnation, Writings of the Ari, Vol. 13, KCI, P. 62
381. Sha'ar Hapsukim, Writings of the Ari, Vol. 8, KCI, P. 188
382. I Samuel 25:3-42
383. I Samuel 25:10
384. Sha'ar Hapsukim, Writings of the Ari, Vol. 8, KCI, P. 188
385. Zechariah 14:9-14:11
386. Ten Luminous Emanations, Vol. 1, Rav Ashlag, KCI, P. 1-2
387. Numbers 14:1-12
388. Numbers 14:29-30
389. Gate of Reincarnation, Writings of the Ari, Vol. 13, KCI
390. Genesis 1:4
391. Kabbalah for the Layman, Vol. 1, Rav Berg, KCI, Ch. 8
392. Isaiah 3:10
393. Talmud Bavli, Tractate Chagigah, P. 12a
394. Genesis 1:4
395. Talmud Bavli, Tractate Ta'anit, P. 61
396. Genesis 1:4
397. Numbers 14:29
398. Numbers 14:2
399. Talmud Bavli, Tractate Ta'anit, P. 29a
400. Ecclesiastes 7:14
401. Genesis 32:25; 32:33
402. Genesis, 25:21-23
403. Zohar, Toldot 4:25
404. Genesis 32:22
405. Genesis 32:25
406. Genesis 32:26
407. Zohar, Vayishlach 7:104
408. Psalms 31:19
409. Zohar, Prologue 6:33
410. Zohar, Prologue 6:33
411. Genesis 3:1-3:25
412. Genesis 3:7
413. Genesis 3:18
414. Ezekiel 47:12
415. The Kabbalah Connection, Rav Berg, KCI, Ch. 15
416. Zohar Hebrew Edition, Prologue, Sulam, par. 33
417. Talmud Bavli, Tractate Gittin, P. 9a
418. Zohar, Prologue 6:33

419. Zohar Hebrew Edition, Prologue, Sulam, par. 33
420. Zohar Hebrew Edition, Prologue, Sulam, par. 33
422. Ecclesiastes 7:14
423. Kabbalah for the Layman, Vol. 1, Rav Berg, KCI, Ch. 11
424. Zohar, Prologue 6:33
425. Talmud Bavli, Tractate Pesachim, P. 56a

Chapter 18

426. Exodus 20:8
427. Zohar, Prologue 6:34
428. Exodus 20:8-11
429. Genesis 2:2-3
430. Exodus 16:22-23
431. The Kabbalah Connection, Rav Berg, KCI
432. Exodus Rabbah, 25:12
433. Talmud Bavli, Tractate Beitzah, P. 16a
434. Genesis Ch. 1
435. Kabbalah for the Layman, Vol. 1, Rav Berg, KCI, Ch. 8
436. Zohar, Beresheet A 30:317
437. Zohar Hebrew Edition, Prologue, Sulam, par. 34
438. The Kabbalah Connection, Rav Berg, KCI, Ch. 15
439. Zohar, Beresheet A, 9:110
440. Zohar. Prologue 6:34

Chapter 19

441. Zohar, Prologue 6:35
442. Exodus 32:19
443. Genesis 2:7
444. Talmud Bavli, Tractate Shabbat, P. 88b
445. Zohar, Beresheet A 22:252-255
446. CF., Ch. 17, P. 21
447. CF., Ch. 17, P. 47.
448. Zohar, Prologue 6:32

449. Zohar Hebrew Edition, Prologue, Sulam, par. 35
450. Zohar Hebrew Edition, Prologue, Sulam, par. 35

Chapter 20

451. Isaiah 45:12
452. Zohar, Vayigash 2:10-11.
453. Deuteronomy 4:24
454. Deuteronomy 4:4
455. Zohar, Beresheet B, 54: 259-262
456. Tikunei Zohar, Prologue, par. 371
457. The Kabbalah Connection, Rav Berg, KCI, Ch. 6
458. Talmud Eser Sefirot (Ten Luminous Emanations), Vol. 1, Rav Ashlag, KCI, P. 31, par. 5-6
459. Psalms 118:19
460. Talmud Bavli, Tractate Shabbat, P. 104a
461. Talmud Eser Sefirot (Ten Luminous Emanations), Vol. 1, Rav Ashlag, KCI, P. 16, par. 8-10
462. Zohar, Prologue 1:1 - 4:13
463. Talmud Eser Sefirot (Ten Luminous Emanations), Vol. 1, Rav Ashlag, KCI
464. Zohar, Lech Lecha 2:4
465. Deuteronomy 15:11
466. Zohar Hebrew Edition, Prologue, par. 36
467. Zohar, Lech Lecha 26:261

Chapter 21

468. Physics and Philosophy, Werner Heisenberg, P. 125
469. Ethics of the Fathers, Ch. 3:1
470. Zohar, Prologue 6:37
471. Zohar, Prologue 6:37

472. Zohar, Emor, 33:214.
473. Genesis 1:27
474. Isaiah 47:13
475. Zohar, Yitro 11:129-130
476. Genesis 1:26
477. Sifrei 11:22
478. Zohar, Shmini 1:3-4
479. The Kabbalah Connection, Rav Berg, KCI, Ch. 18
480. Zohar, Vayelech 7:39
481. Zohar, Shemini 1:5-6
482. Exodus 14:19-14:21
483. Zohar, Beshalach 14:160-184.
484. Malachi 3:10
485. Zohar Hebrew Edition, Prologue, Sulam, par. 37
486. Genesis 9:26-9:27
487. Genesis Ch. 27; Ch. 28, 1-4
488. Genesis Ch. 49
489. Genesis 48:13-48:22
490. Genesis 48:20
491. Numbers 6:24-6:26
492. Malachi 2:7
493. Zohar, Naso 16:137-138
494. Psalms 89:3
495. Zohar Hebrew Edition, Prologue, Sulam, par.37
496. Kabbalah for the Layman, Vol. 1, Rav Berg, KCI, Ch. 8 and 9
497. A Study of History, Toynbee
498. Ecclesiastes 1:9
499. Zohar Hebrew Edition, Prologue, Sulam, par. 37
500. Malachi 3:6
501. Deuteronomy 8:10
502. Psalms 145:16
503. Zohar, Terumah 46:499
504. Kabbalah for the Layman, Vol. 1, Rav Berg, KCI, Ch. 8 and
505. Zohar, Terumah 46:499 9
506. Zohar, Ekev 1:1-10.
507. Zohar Hebrew Edition, Prologue, Sulam, par. 37
508. Psalms 22:20
509. Zohar, Tetzaveh 1:6
510. Zohar Chadash, Midrash Eicha, Sulam, par. 108
511. Exodus, Ch. 32
512. Ecclesiastes 10:2
513. Zohar, Beresheet A, 22:255
514. Exodus 32:19.
515. Zohar, Beresheet A, 22:255
516. Zohar Hebrew Edition, Vayera, Sulam, par. 48-49.
517. Malachi 3:18
518. Zohar Hebrew Edition, Prologue, Sulam, par. 37.

Chapter 22

519. Zohar, Prologue 6:38
520. Sefer Yetzirah (Book of Formation), P. 47a
521. Kabbalah for the Layman, Vol. 1, Rav Berg, KCI, Ch. 8
522. Introduction to the Zohar, Rav Ashlag, KCI, Par. 34-36
523. Zohar, Vayikra 10:75-80
524. Amos 5:2
525. Zohar, Vayikra 10:78-81
526. Amos 9:11
527. Zohar Hebrew Edition, Prologue, Sulam, par. 38

THE ZOHAR

Composed more than 2,000 years ago, the *Zohar* is a set of 23 books, a commentary on biblical and spiritual matters in the form of conversations among spiritual masters. But to describe the *Zohar* only in physical terms is greatly misleading. In truth, the *Zohar* is nothing less than a powerful tool for achieving the most important purposes of our lives. It was given to all humankind by the Creator to bring us protection, to connect us with the Creator's Light, and ultimately to fulfill our birthright of true spiritual transformation.

More than eighty years ago, when The Kabbalah Centre was founded, the *Zohar* had virtually disappeared from the world. Few people in the general population had ever heard of it. Whoever sought to read it—in any country, in any language, at any price— faced a long and futile search.

Today all this has changed. Through the work of The Kabbalah Centre and the editorial efforts of Michael Berg, the *Zohar* is now being brought to the world, not only in the original Aramaic language but also in English. The new English *Zohar* provides everything for connecting to this sacred text on all levels: the original Aramaic text for scanning; an English translation; and clear, concise commentary for study and learning.

MORE BOOKS THAT CAN HELP YOU BRING THE WISDOM OF KABBALAH INTO YOUR LIFE

Nano: Technology of Mind over Matter
By Rav Berg

Kabbalah is all about attaining control over the physical world, including our personal lives, at the most fundamental level of reality. It's about achieving and extending mind over matter and developing the ability to create fulfillment, joy, and happiness by controlling everything at the most basic level of existence. In this way, Kabbalah predates and presages the most exciting trend in recent scientific and technological development, the application of nanotechnology to all areas of life in order to create better, stronger, and more efficient results.

Immortality: The Inevitability of Eternal Life
By Rav Berg

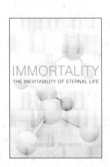

This book will totally change the way in which you perceive the world, if you simply approach its contents with an open mind and an open heart.

Most people have it backwards, dreading and battling what they see as the inevitability of aging and death. But,

439

according to the great Kabbalist Rav Berg and the ancient wisdom of Kabbalah, it is eternal life that is inevitable.

With a radical shift in our cosmic awareness and the transformation of the collective consciousness that will follow, we can bring about the demise of the death force once and for all— in this "lifetime."

Simple Light
By Karen Berg

From the woman regarded by many as their "spiritual mother," and whose work has touched millions of lives around the world, here is a book with a message that is simple and straight from the heart: It's all about love and sharing.

Karen's unique voice will serve to inspire you and help you to face life's daily challenges. Open the book to any passage whenever you find a moment, and you will begin to discover the keys to leading a more joyful and fulfilled life.

The Power of Kabbalah
By Yehuda Berg

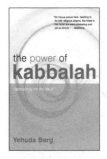

Imagine your life filled with unending joy, purpose, and contentment. Imagine your days infused with pure insight and energy. This is The Power of Kabbalah. It is the path from the momentary pleasure that most of us settle for, to the lasting fulfillment that is yours to claim. Your deepest desires are waiting to be realized. Find out how, in this basic introduction to the ancient wisdom of Kabbalah.

Secrets of the Zohar: Stories and Meditations to Awaken the Heart
By Michael Berg

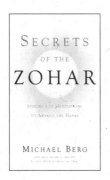

The *Zohar*'s secrets are the secrets of the Bible, passed on as oral tradition and then recorded as a sacred text that remained hidden for thousands of years. They have never been revealed quite as they are here in these pages, which decipher the codes behind the best stories of the ancient sages and offer a special meditation for each one. Entire portions of the *Zohar* are presented, with the Aramaic and its English translation in side-by-side columns. This allows you to scan and to read aloud so that you can draw on the *Zohar*'s full energy and achieve spiritual transformation. Open this book and open your heart to the Light of the *Zohar*!

THE KABBALAH CENTRE

The Kabbalah Centre is a spiritual organization dedicated to bringing the wisdom of Kabbalah to the world. The Kabbalah Centre itself has existed for more than 80 years, but its spiritual lineage extends back to Rav Isaac Luria in the 16th century and even further back to Rav Shimon bar Yochai, who revealed the principal text of Kabbalah, the *Zohar*, more than 2,000 years ago.

The Kabbalah Centre was founded in 1922 by Rav Yehuda Ashlag, one of the greatest kabbalists of the 20th Century. When Rav Ashlag left this world, leadership of The Kabbalah Centre was taken on by Rav Yehuda Brandwein. Before his passing, Rav Brandwein designated Rav Berg as director of The Kabbalah Centre. Now, for more than 30 years, The Kabbalah Centre has been under the direction of Rav Berg, his wife Karen Berg, and their sons, Yehuda Berg and Michael Berg.

Although there are many scholarly studies of Kabbalah, The Kabbalah Centre does not teach Kabbalah as an academic discipline but as a way of creating a better life. The mission of The Kabbalah Centre is to make the practical tools and spiritual teachings of Kabbalah available to everyone.

The Kabbalah Centre makes no promises. But if people are willing to work hard to grow and become actively sharing, caring and tolerant human beings, Kabbalah teaches that they will then experience fulfillment and joy in a way previously unknown to them. This sense of fulfillment, however, comes gradually and is always the result of the student's spiritual work.

Our ultimate goal is for all humanity to gain the happiness and fulfillment that is our true destiny.

Kabbalah teaches its students to question and test everything they learn. One of the most important teachings of Kabbalah is that there is no coercion in spirituality.

What Does The Kabbalah Centre Offer?

Local Kabbalah Centres around the world offer onsite spiritual services, lectures, classes, study groups, holiday celebrations and services, and a community of teachers and fellow students. To find a Centre near you, go to www.kabbalah.com.

For those of you unable to access a physical Kabbalah Centre due to the constraints of location or time, we have other ways to participate in The Kabbalah Centre community.

At www.kabbalah.com, we feature online blogs, newsletters, weekly wisdom, a store, and much more.

It's a wonderful way to stay tuned in and in touch, and it gives you access to programs that will expand your mind and challenge you to continue your spiritual work.

Student Support

The Kabbalah Centre empowers people to take responsibility for their own lives. It's about the teachings, not the teachers. But on your journey to personal growth, things can be unclear and sometimes rocky, so it is helpful to have a coach or teacher. Simply call 1 800 KABBALAH toll free.

All Student Support instructors have studied Kabbalah under the direct supervision of Kabbalist Rav Berg, widely recognized as the preeminent kabbalist of our time.

We have also created opportunities for you to interact with other Student Support students through study groups, monthly connections, holiday retreats, and other spiritual events held around the country.

KABBALAH UNIVERSITY
FOUNDED UPON THE TEACHINGS OF RAV BERG

Be in the center of Kabbalah activities anytime and anywhere through ukabbalah.com

Kabbalah University (www.ukabbalah.com) is an online resource center and community offering a vault of wisdom spanning 30 years, and rapidly growing. Removing any time-space limitation, this virtual Kabbalah Centre presents the same courses and spiritual connections as the physical centers, with an added benefit of live streaming videos from worldwide travels. As close as a click of your finger, for a low monthly access fee, it's open 24/7.

Stay current with historic lessons from Rav Berg and inspiring talks with Karen. Delve deeper into Michael Berg's teachings, and journey with Yehuda Berg to holy sites. Connect with world-renowned Kabbalah instructors sharing weekly *Zohar* and consciousness classes that awaken insights into essential life matters such as: relationships, health, prosperity, reincarnation, parenting, and astrology. Check out the library, including hundreds of spiritual topics going back more than four decades. A richer world awaits your presence at ukabbalah.com.

With deepest love and gratitude to my parents
Moshe ben Mazal and Aziza bat Naeema
for everything you have given our entire family.

May we all merit through this dedication and
spreading the wisdom of Kabbalah to be protected
and connected to the Tree of Life.

And for all those who are seeking their soul mates,
may we all be blessed with uniting with them and
sharing a spiritual life together.

Sarah